3rd International Workshop on New Computational Methods for Inverse Problems

(NCMIP 2013)

Journal of Physics: Conference Series Volume 464

Cachan, France
22 May 2013

ISBN: 978-1-62993-283-5
ISSN: 1742-6588

Printed from e-media with permission by:

Curran Associates, Inc.
57 Morehouse Lane
Red Hook, NY 12571

CURRAN ASSOCIATES INC. proceedings.com

Some format issues inherent in the e-media version may also appear in this print version.

Printed by Curran Associates, Inc. (2013)

For permission requests, please contact the Institute of Physics at the address below.

Institute of Physics
Dirac House, Temple Back
Bristol BS1 6BE UK

Phone: 44 1 17 929 7481
Fax: 44 1 17 920 0979

techtracking@iop.org

Additional copies of this publication are available from:

Curran Associates, Inc.
57 Morehouse Lane
Red Hook, NY 12571 USA
Phone: 845-758-0400
Fax: 845-758-2634
Email: curran@proceedings.com
Web: www.proceedings.com

TABLE OF CONTENTS

Journal of Physics: Conference Series **464** (2013) 011001 doi:10.1088/1742-6596/464/1/011001

Foreword

This volume of *Journal of Physics: Conference Series* is dedicated to the scientific contributions presented during the 3rd International Workshop on New Computational Methods for Inverse Problems, NCMIP 2013 (http://www.farman.ens-cachan.fr/NCMIP_2013.html). This workshop took place at Ecole Normale Supérieure de Cachan, in Cachan, France, on 22 May 2013, at the initiative of Institut Farman. The prior editions of NCMIP also took place in Cachan, France, firstly within the scope of the ValueTools Conference, in May 2011 (http://www.ncmip.org/2011/), and secondly at the initiative of Institut Farman, in May 2012 (http://www.farman.ens-cachan.fr/NCMIP_2012.html).

The NCMIP Workshop focused on recent advances in the resolution of inverse problems. Indeed inverse problems appear in numerous scientific areas such as geophysics, biological and medical imaging, material and structure characterization, electrical, mechanical and civil engineering, and finances. The resolution of inverse problems consists of estimating the parameters of the observed system or structure from data collected by an instrumental sensing or imaging device. Its success firstly requires the collection of relevant observation data. It also requires accurate models describing the physical interactions between the instrumental device and the observed system, as well as the intrinsic properties of the solution itself. Finally, it requires the design of robust, accurate and efficient inversion algorithms. Advanced sensor arrays and imaging devices provide high rate and high volume data; in this context, the efficient resolution of the inverse problem requires the joint development of new models and inversion methods, taking computational and implementation aspects into account. During this one-day workshop, researchers had the opportunity to bring to light and share new techniques and results in the field of inverse problems.

The topics of the workshop were: algorithms and computational aspects of inversion, Bayesian estimation, kernel methods, learning methods, convex optimization, free discontinuity problems, metamodels, proper orthogonal decomposition, reduced models for the inversion, non-linear inverse scattering, image reconstruction and restoration, and applications (bio-medical imaging, non-destructive evaluation...).

NCMIP 2013 was a one-day workshop held in May 2013 which attracted around 60 attendees. Each of the submitted papers has been reviewed by three reviewers. Among the accepted papers, there are seven oral presentations, five posters and one invited poster (On a deconvolution challenge presented by C Vonesch from EPFL, Switzerland). In addition, three international speakers were invited to present a longer talk.

The workshop was supported by Institut Farman (ENS Cachan, CNRS) and endorsed by the following French research networks (GDR ISIS, GDR Ondes, GDR MOA, GDR MSPC). The program committee acknowledges the following research laboratories CMLA, LMT, LSV, LURPA, SATIE.

Laure Blanc-Féraud and Pierre-Yves Joubert

Workshop co-chair
Laure Blanc-Féraud, I3S laboratory and INRIA Nice Sophia-Antipolis, France
Pierre-Yves Joubert, IEF, Paris-Sud University, CNRS, France

Technical program committee
Gilles Aubert, J-A Dieudonné Laboratory, CNRS and University of Nice-Sophia Antipolis, France
Nabil Anwer, LURPA, ENS Cachan, France
Alexandre Baussard, ENSTA Bretagne, Lab-STICC, France
Marc Bonnet, ENSTA, ParisTech, France
Antonin Chambolle, CMAP, Ecole Polytechnique, CNRS, France
Oliver Dorn, School of Mathematics, University of Manchester, UK
Cécile Durieu, SATIE, ENS Cachan, CNRS, France
Gérard Favier, I3S Laboratory, University of Nice Sophia-Antipolis, France
Mário Figueiredo, Instituto Superior Técnico, Lisbon, Portugal
Laurent Fribourg, LSV, ENS Cachan, CNRS, France
Marc Lambert, L2S Laboratory, CNRS, SupElec, Paris-Sud University, France
Dominique Lesselier, L2S Laboratory, CNRS, SupElec, Paris-Sud University, France
Matteo Pastorino, DIBE, University of Genoa, Italy
Christian Rey, LMT, ENS Cachan, CNRS, France
Simon Setzer, Saarland University, Germany
Cedric Vonesch, EPFL, Switzerland

Local chair
Sophie Abriet, SATIE Laboratory, ENS Cachan, France
Béatrice Bacquet, SATIE Laboratory, ENS Cachan, France
Lydia Matijevic, LMT Laboratory, ENS Cachan France

Invited speakers
Jérôme Idier, IRCCyN (UMR CNRS 6597), Ecole Centrale de Nantes, France
Massimo Fornasier, Faculty of Mathematics, Technical University of Munich, Germany
Matthias Fink, Institut Langevin, ESPCI, Université Paris Diderot, France

Journal of Physics: Conference Series **464** (2013) 011002 doi:10.1088/1742-6596/464/1/011002

Peer review statement

All papers published in this volume of *Journal of Physics: Conference Series* have been peer reviewed through processes administered by the proceedings Editors. Reviews were conducted by expert referees to the professional and scientific standards expected of a proceedings journal published by IOP Publishing.

Journal of Physics: Conference Series **464** (2013) 012001 doi:10.1088/1742-6596/464/1/012001

A time-reversal process for beam trusses subjected to impulse loads

Yves Le Guennec
Laboratoire MSSMat, ECP – F-92295 Châtenay-Malabry, France
E-mail: yves.le-guennec@ecp.fr

Éric Savin
DADS, ONERA – The French Aerospace Lab – F-92322 Châtillon cedex, France
E-mail: eric.savin@onera.fr

Didier Clouteau
Laboratoire MSSMat, ECP – F-92295 Châtenay-Malabry, France
E-mail: didier.clouteau@ecp.fr

Abstract. Spatial structures are often subjected to impulse loads which give rise to high-frequency (HF) waves. The objectives of the research outlined in this paper are twofold: (i) develop a reliable *direct* model of the transient dynamic response of built-up structures subjected to such impulse loads and (ii) use this model in a time-reversal *inverse* process in order to possibly detect the location of the shocks. The present study is more particularly focused on beam assemblies, as typically encountered in aerospace applications. At first, a transport model describing the evolution of the HF vibrational energy density in a three-dimensional beam truss is outlined, with an emphasis on the consideration of the reflection/transmission phenomena at the beam junctions. The latter contribute to spread HF waves within the entire structure, yielding a diffuse (noisy) vibrational energy field amenable to be efficiently time-reversed for the purpose of locating the source which has generated it. The time-reversal process itself is presented in a subsequent part. Its originality lies in the consideration of quadratic observables (energy densities) as the processed data. The approach is illustrated by a numerical simulation performed on a three-dimensional beam truss.

1. Introduction

Spatial structures are often subjected to impulse loads due to planned processes–like pyrotechnic shocks for solar panel unfolding on a satellite structure–or accidental events–like mechanical shocks generated by the collision with an orbital waste. Such loads propagate high-frequency (HF) waves carrying a vibrational energy that may be harmful to the structures themselves, or may disturb the functioning of the equipments attached to them [1]. Engineers have to resort to transient models in order to simulate and predict the response of mechanical systems to shock excitations. These analyses are still a challenging issue, both from a theoretical and a numerical point of view. In addition, a knowledge of the energy paths may be desirable for the design of aerospace devices or for the localization of impacts on structures. This paper focuses on an approach developed to locate impulse loads in beam trusses based on the measurement of quadratic observables (vibrational energies), and the use of a time-reversal process [2] applied to these signals.

Several classes of methods have been developed by engineers in order to predict the dynamic response of built-up structures to impact loads. Wave tracking approaches have been proposed such as the WKBJ (for Wentzel, Kramers, Brillouin, and Jeffreys) method [3] or the segment projection method [4], which are both dedicated to the HF range. Nevertheless as the number of wave reflections and transmissions at boundaries or interfaces between substructures increases, the number of rays to be tracked increases as well, leading to prohibitive numerical costs. Another class of local approach called the Vibrational Conductivity Analogy (VCA) has also been introduced [5]. VCA is based on the assumption that a diffusion-like evolution equation holds for the HF wave energy density. This is contradictory because a diffusion equation is not reversible in time, while this model arises from a wave equation which is reversible in time. One advantage of VCA is the consideration of energy densities, though. Indeed, these quantities evolve in the HF range at a typical scale comparable to the dimensions of the substructures in a built-up system [6]. This allows for numerical resolutions at a reduced numerical cost because the underlying (possibly sharp) wave fronts need not to be tracked. The model retained in this research aims to take advantage of both approaches: lower numerical costs achieved by the use of quadratic observables (energy densities) as in the VCA, while satisfying the time reversibility property of waves.

Based on a microlocal analysis of classical wave systems, it can be proved that the energy density associated with their HF solutions satisfies a Liouville-type transport equation [7]. The latter is time reversible as imposed by wave physics, and describes energy fields which are smoother than the underlying wave fields. Thus they are easier to solve numerically. Here, this so-called kinetic model is adapted to the case of three-dimensional beam trusses as done in [8]. The transport equations are solved numerically using modal or nodal discontinuous Galerkin (DG) finite elements [9] together with a strong-stability preserving (SSP) Runge-Kutta scheme [10] for time integration. DG methods have the advantage to be weakly dissipative and weakly dispersive, depending on the order of the nodal interpolation or the modal expansion. Thus they are well suited for late-time simulations. This is a much desirable feature for time-reversal experiments because time reversal performs better in cluttered media using multiply scattered waves recorded over long periods. At long times, the multiple scattering of HF waves induced by their multiple reflections and transmissions at the junctions and interfaces of a built-up structure are well described by a kinetic model in terms of the associated energy density [6]. Our objective is then to use this multiply scattered scalar field as the signal to be time reversed for the purpose of locating the source (the shock) which has generated the underlying HF waves. Thus the overall strategy of this research is the following: (i) one develops a kinetic model of HF wave propagation phenomena in terms of smoother quadratic observables, namely the associated energy densities; (ii) these data are used in time-reversal experiments in order to locate their source, after they have been multiply scattered by the beam junctions in the present case.

The paper is organized as follows. In a first section, the kinetic model for the evolution of wave energy densities is outlined, and the reflection/transmission phenomena pertaining to built-up structures are considered for the case of point junctions of several beams in a beam truss. The corresponding equations to be considered for time reversed simulations are detailed in the section 3, paying a particular attention to the treatment of the reflection/transmission phenomena in the time-reversed process. The numerical implementation and some preliminary numerical results are discussed in the section 4. The last section offers some conclusions and perspectives.

Journal of Physics: Conference Series **464** (2013) 012001 doi:10.1088/1742-6596/464/1/012001

2. Kinetic model for beam trusses using Timoshenko's kinematics

The purpose of this section is to summarize the aforementioned kinetic model of HF wave propagation in a single Timoshenko beam and the reflection/transmission phenomena arising at the junctions of several beams, in order to consider complex three-dimensional beam trusses. The vibrational behavior of the beam is modeled by Timoshenko's beam kinematics [11] because it is more suitable to the HF range than the Euler-Bernoulli one. It includes the effects of rotary inertia and shear distortion. In this paper beam's material is assumed to be isotropic and its heterogeneities vary slowly in regard to the considered wavelengths. The cross-sections are assumed to be symmetric such that their geometrical and inertial centers are coincident and constitute the neutral fiber of the beam. The orthonormal axes attached to the neutral fiber will be denoted by $(\hat{\mathbf{e}}_1, \hat{\mathbf{e}}_2, \hat{\mathbf{e}}_3)$, $\hat{\mathbf{e}}_1$ being the tangent axis, and the corresponding coordinates by x, y, z. By Timoshenko's beam kinematical assumption, the displacement of a point $\boldsymbol{x}$ of the cross-section is split up into the contribution of the translation of the neutral fiber and the contribution of the rotation of the cross-section. Hence the motion of each cross-section is described by six degrees of freedom, depending on the tangent (curvilinear) coordinate x solely. In this framework, it is possible to show that the energy density $w_\alpha(x, t, k)$ of a HF wave propagating in the beam in the polarization mode α with wave vector k is the solution of the transport equation [6, 7]:

$$\partial_t w_\alpha + c_\alpha \hat{k}\, \partial_x w_\alpha = 0\,, \quad \alpha \in \mathsf{E}\,, \tag{1}$$

where $|k|$ is the wavenumber, $\hat{k} = k/|k|$ is the sign of the wavenumber, c_α is the group velocity of the mode α, and $\mathsf{E} = \{\mathrm{P}_1, \mathrm{P}_2, \mathrm{P}_3, \mathrm{T}_1, \mathrm{T}_2, \mathrm{T}_3\}$ is the set of the six possible polarizations. P_i denotes the compressional modes and T_i denotes the transverse modes for $1 \leq i \leq 3$. The group velocities are given by $c_\mathrm{P}(x) = \sqrt{E/\rho}$ for the compressional energetic waves and $c_\mathrm{T}(x) = \sqrt{\kappa\mu/\rho}$ for the transverse energetic waves, where E is the Young's modulus of the considered beam, $\kappa\mu$ is its reduced shear modulus, and ρ is its material density. One remarks that the k-dependency of the energy density w_α allows for wave crossing and superposition. The ultimate quantity to consider is the overall space-time energy density given by $\mathcal{E}(x,t) = \sum_{\alpha\in\mathsf{E}} \int_\mathbb{R} w_\alpha(x,k,t)\, dk$.

Now the propagation of the HF vibrational energy density in a single beam being described by Eqs. (1), the reflection/transmission phenomena arising at the junctions have to be considered in order to obtain the response of a full beam assembly. They contribute to couple the energy densities w_α that are partially reflected and transmitted at the junctions. These reflections and transmissions will give rise to a diffuse energy field spread over the entire truss at late times. It constitutes the noisy data used in the time-reversal process proposed in this paper. These phenomena are represented by power flow reflection/transmission coefficients for a junction of several of beams:

$$\rho_{\alpha\beta}^{11} = \frac{< \Pi_\alpha^R >}{< \Pi_\beta^I >}\,, \quad \tau_{\alpha\beta}^{q1} = \frac{< \Pi_\alpha^{Tq} >}{< \Pi_\beta^I >}\,, \tag{2}$$

where $< \Pi_\alpha^R >$ is the time average power flow of the reflected waves with the polarization $\alpha \in \mathsf{E}$, $< \Pi_\alpha^{Tq} >$ is the time average power flow of the transmitted waves with the polarization α in beam $\#q$, and $< \Pi_\beta^I >$ is the time average power flow of an incident wave with the polarization β. The time average power flows are computed from the continuity of the displacements, rotations, forces and moments at the junction [8]. It should be noted that the transport equations provided with boundary conditions corresponding to the above reflection/transmission operators and some initial conditions (the impulse loads one ultimately wish to localize) form a well-posed system.

3. Numerical time-reversal process

The aim of a time-reversal experiment is to possibly reconstruct an initial pulse from time-reversed signals. In the original setting of [2], the time-reversed signals were amplified beforehand in order to reconstruct an amplified pulse able to break up kidney stones, for example. Here the

Journal of Physics: Conference Series **464** (2013) 012001 doi:10.1088/1742-6596/464/1/012001

time-reversal procedure is considered for the transport equation (1) and reflection/transmission boundary conditions (2). The time-reversed energy density field $w_\alpha^-(x,k,t)$ should have the property:

$$w_\alpha^-(x,k,t) = w_\alpha(x,-k,t_{\text{end}}-t)\,, \tag{3}$$

meaning that the energy density traveling in a direction $\hat{k}$ at time t and position x in the direct configuration travels in a direction $-\hat{k}$ at time $t_{\text{end}} - t$ and position x in the reversed configuration. Now considering at first an isolated beam, the time-reversed energy density field within that beam is obtained straightforwardly: as the evolution law (1) is reversible in time, the time-reversed evolution equation is the same.

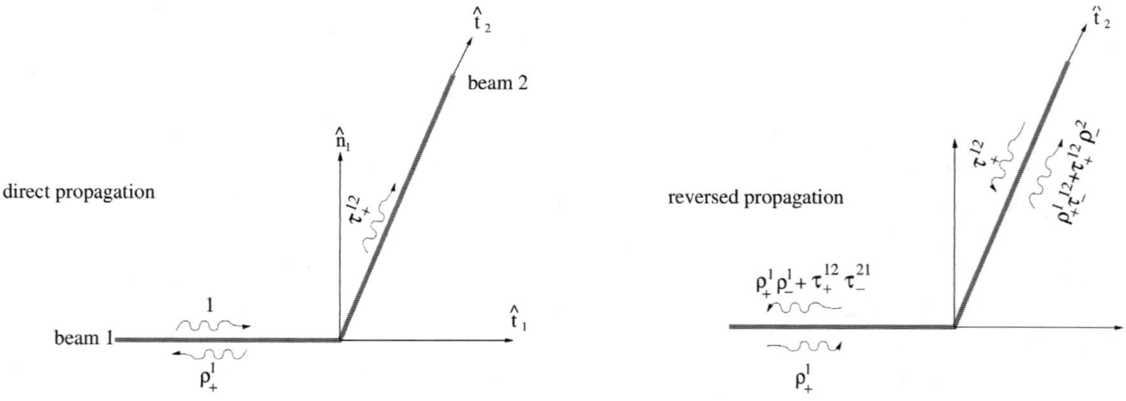

Figure 1: Energy waves in the direct propagation and their compatibility conditions for a time-reversed propagation. The wavy arrows indicate propagation directions.

The reversed-time reflection/transmission operators at the junctions have also to be derived. As shown in Figure 1, in the direct configuration, an incident energy flow normalized to 1 in the mode α generates reflected and transmitted flows of amplitudes given by the reflection/transmission coefficients given by Eqs. (2) when it impinges the junction. In the time-reversed configuration the wavenumber, and so the propagation direction, are the opposite of their direct counterparts. Thus the impinging flows are now the reflected and transmitted flows once they have been reversed in time; see Figure 1. When impinging the junction, these flows generate reflected and transmitted flows themselves, according to some reversed-time reflection/transmission operators. So the latter should be constructed such that one recovers the initial flow in mode α from the reflected/transmitted flows. These compatibility conditions yields the matrix formulation:

$$\begin{bmatrix} \boldsymbol{\rho}_+^1 & \boldsymbol{\tau}_+^{12} \\ \boldsymbol{\tau}_+^{21} & \boldsymbol{\rho}_+^2 \end{bmatrix} \begin{bmatrix} \boldsymbol{\rho}_-^1 & \boldsymbol{\tau}_-^{12} \\ \boldsymbol{\tau}_-^{21} & \boldsymbol{\rho}_-^2 \end{bmatrix} = \boldsymbol{I}\,, \tag{4}$$

where $\boldsymbol{\rho}_-^p, \boldsymbol{\tau}_-^{pq}$ are the reversed-time reflection/transmission matrices from beam $\#p$ to beam $\#q$. These operators are the unknowns of the system (4), while $\boldsymbol{\rho}_+^p$ and $\boldsymbol{\tau}_+^{pq}$ are the reflection and transmission operators for direct propagation as defined in Eq. (2). Thus it appears that the reversed-time reflection/transmission operators for the junction are simply the inverse of the direct reflection/transmission operators.

4. Numerical example

Because we lack experimental data, the signals to be time reversed are the energy density fields simulated by a direct computation up to a final time t_{end}. The transport equations are solved numerically using a discontinuous Galerkin (DG) scheme for spatial discretization [9] and an SSP Runge-Kutta scheme for time integration [10]. This numerical scheme is detailed in [8]. The goal of the simulation presented here is to test it for its use with the time-reversal process sketched in the foregoing section. Provided it is weakly dispersive and weakly dissipative, then it should be possible to retrieve an initial pulse from time-reversed simulated signals up to late times. The time-reversed simulation shall be performed using the direct energy density field computed at t_{end} as the initial pulse for that configuration, considering however an opposite direction of propagation. If the aforementioned properties of the numerical scheme are confirmed, then one argues that it can be used in a real time-reversal procedure using real data recorded from a real structure. A similar approach has been adopted in [12] for example, in order to test the time-reversal capability of a spectral finite element code solving the elastic wave equation in an unbounded, three-dimensional anisotropic random medium.

The numerical time-reversal procedure has been applied to the truss of Figure 2. The Young's moduli and beam's lengths are gathered in Tab. 1, while the Poisson's ratio is set to 0.3 for the entire truss. The direct simulation is performed up to $t_{\text{end}} = 2 \times T$, where T is the time required by a transverse wave to travel across the first beam. Figure 3 shows the evolution of the energy density in the truss at selected times, and the evolution of the total energies in each beam, respectively, up to $t = 4 \times T$ after time reversal has been performed. It shows that the energy is rather uniformly spread over the entire truss at $t = t_{\text{end}}$. All these plots are symmetric about t_{end}. This feature underlines the weak dissipation and weak dispersion of the numerical scheme because the evolution of the energy is symmetric for the direct and reversed simulations. Would the scheme lack these properties, it would be impossible to observe any symmetry because numerical dispersion and dissipation would have modified the signals too much to be able to retrieve the initial pulse. Here that initial pulse is indeed well retrieved.

beam $\#p$	length L_p	Young's modulus
1	5	1
2	4	0.75
3	5	1.2
4	5	1
5	2.5217	1.25
6	4	1
7	3.6988	1.25
8	5.6853	0.5
9	6.4031	1.5

Table 1: Normalized Young's moduli and beam lengths for the truss ofFigure 2.

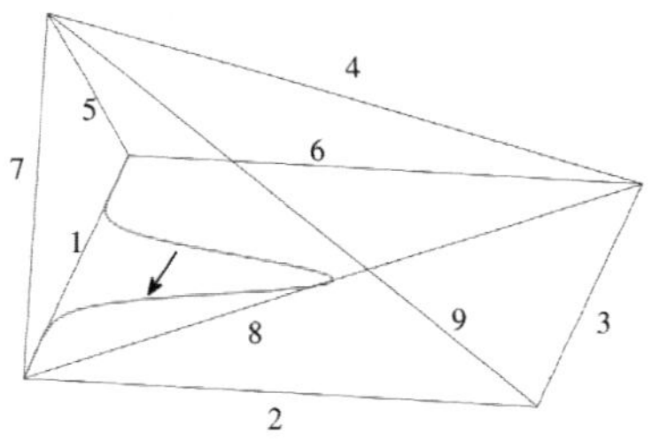

Figure 2: View of the single-bay beam truss with a Gaussian-shaped initial load in beam #1 (red line).

5. Conclusions

A kinetic model for the evolution of the vibrational energy density in Timoshenko beam assemblies has been outlined. The reflection/transmission phenomena at the beam junctions are taken into account through power flow reflection/transmission coefficients. They allow to consider complex three-dimensional beam trusses. The advantage of the model and the proposed numerical scheme is the possibility to perform numerical time-reversal experiments with quadratic observables. A numerical example has been presented showing the feasibility of this approach. Now this numerical test only constitutes the first step of a complete time-reversal

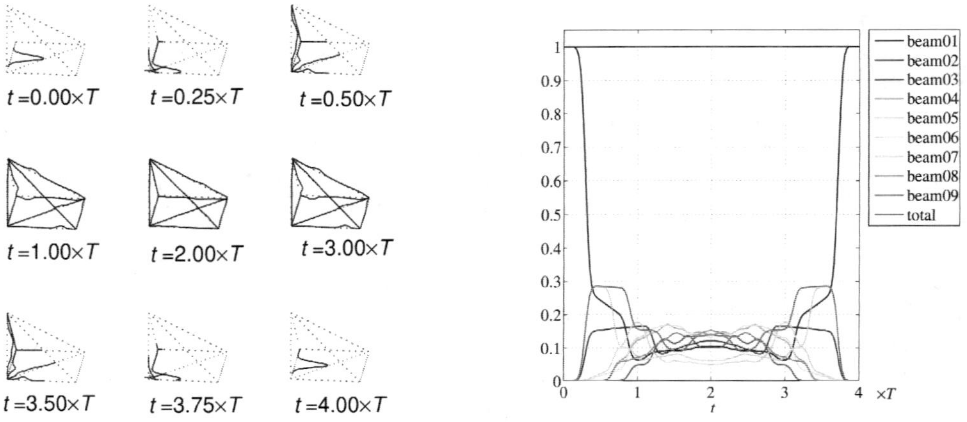

(a) Energy density evolution (b) Integrated energy density

Figure 3: (a) Direct and reversed evolution of the energy density within the truss of Figure 2 at selected times for a compressional initial pulse. Time reversal is performed at $t = 2 \times T$. (b) Evolution of the total energies $\mathcal{E}^p(t)$, $1 \leq p \leq 9$, in each beam of the truss of Figure 2.

procedure. The next step should be a stability study of the numerical scheme in order to estimate the maximum time t_{end} at which the initial pulse is still retrieved. The analysis of the effect of some noise on the numerical scheme should be performed in view of developing laboratory experiments, and time-reversal mirrors could be used subsequently. The data processed by the latter shall be the overall history of the energy density at selected points instead of the energy density recorded on the entire structure at a final time. This option has already been tested and gives satisfactory results. It will be reported in future communications.

References

[1] C. J. Moening 1986, Views of the world of pyrotechnic shock, *Shock and Vibration Bulletin*, **56**:3-28.

[2] M. Fink 1997, Time reversed acoustics, *Physics Today*, **50**:34-40.

[3] C. R. Steele 1969, Asymptotic analysis of stress waves in inhomogeneous elastic solids, *Journal of American Institute of Aeronautics and Astronautics*, **7**:896-902.

[4] B. Engquist; O. Runborg; A.-K. Tornberg 2002, High-frequency wave propagation by the segment projection method, *Journal of Computational Physics*, **178**:373-390.

[5] D. J. Nefske; S. H. Sung 1989, Power flow finite element analysis of dynamic systems: basic theory and application to beams, *ASME Journal of Vibration, Acoustics, Stress and Reliability in Design*, **111**:94-100.

[6] É. Savin 2013, High-frequency dynamics of heterogeneous slender structures, *Journal of Sound and Vibration*, **332**:2461-2487.

[7] L. Ryzhik; G. Papanicolaou; J. B. Keller 1996, Transport equations for elastic and other waves in random media, *Wave Motion*, **24**:327-370.

[8] Y. Le Guennec; É. Savin 2011, A transport model and numerical simulation of the high-frequency dynamics of three-dimensional beam trusses, *Journal of the Acoustical Society of America*, **130**:3706-3722.

[9] J. S. Hesthaven; T. Warburton 2008, *Nodal Discountinuous Galerkin Methods*, Springer.

[10] S. Gottlieb; C.-W. Shu; E. Tadmor 2001, Strong stability-preserving high-order time discretization methods, *SIAM Review*, **43**:89-112.

[11] S. P. Timoshenko 1922, On the transverse vibrations of bars of uniform cross-section, *Philosophical Magazine*, **43**:125-131.

[12] Q.-A. Ta; D. Clouteau; R. Cottereau 2010, Modeling of random anisotropic elastic media and impact on wave propagation, *European Journal of Computational Mechanics*, **19**:241-253.

Journal of Physics: Conference Series **464** (2013) 012002 doi:10.1088/1742-6596/464/1/012002

Coexistence of Near-Field and Far-Field Sources: the Angular Resolution Limit

Rémy Boyer[♮], Mohammed Nabil El Korso[♯], Alexandre Renaux[♮] and Sylvie Marcos[♮]

[♮]Laboratoire des Signaux et Systemes (L2S)
Université Paris-Sud XI (UPS), CNRS, SUPELEC,
Gif-Sur-Yvette, France

[♯]Laboratoire SATIE, CNRS
École Normale Suprieure de Cachan
Cachan, France

E-mail: `remy.boyer@lss.supelec.fr`

Abstract. Passive source localization is a well known inverse problem in which we convert the observed measurements into information about the direction of arrivals. In this paper we focus on the optimal resolution of such problem. More precisely, we propose in this contribution to derive and analyze the Angular Resolution Limit (ARL) for the scenario of mixed Near-Field (NF) and Far-Field (FF) Sources. This scenario is relevant to some realistic situations. We base our analysis on the Smith's equation which involves the Cramér-Rao Bound (CRB). This equation provides the theoretical ARL which is independent of a specific estimator. Our methodology is the following: first, we derive a closed-form expression of the CRB for the considered problem. Using these expressions, we can rewrite the Smith's equation as a 4-th order polynomial by assuming a small separation of the sources. Finally, we derive in closed-form the analytic ARL under or not the assumption of low noise variance. The obtained expression is compact and can provide useful qualitative informations on the behavior of the ARL.

1. Introduction

Very few works are related to the study of the realistic situation where there exists coexisting far-field (FF) and near-field (NF) sources [2] such as speaker localization using microphone arrays and guidance (homing) systems. At the contrary, we can find a plethora of contributions on the localization of far-field sources [1]. More recently, the problem of localization of near-field sources has been tackled in reference [4–6, 11] for instance. In the context of the problem of source localization, one can see three contributions: (1) propose new efficient algorithms/estimators [1], (2) study the estimation performance independently of a specific algorithm thanks to the lower bound on the Mean Square error (MSE) [3,5] and (3) derive and study the theoretical resolution, *i.e.*, the minimal angular distance to resolve/discriminate two closely spaced emitted signals in terms of their direction of arrivals. Our contribution belongs to the third point. More precisely, based on the Smith's equation [7, 9, 10] which involves the Cramér-Rao Bound (CRB) [3, 12], we derive and analyze the Angular Resolution Limit (ARL) for the realistic scenario where we have two sources, one located in the far-field of the array and another in the near-field of the array.

2. Model setup

We consider some practical applications where the signals $s_1(t)$ and $s_2(t)$ with $t \in [1 : T]$ in which T is the number of snapshots received by an uniform linear array composed by L sensors are the mixture of near-field and far-field sources. More precisely, the $(TL) \times 1$ observation vector is defined as $\mathbf{y} = \bar{\mathbf{y}} + \mathbf{e}$ where $\mathbf{e}$ is the complex centered circular additive white Gaussian noise of variance σ^2 and the $(TL) \times 1$ noise-free signal is $\bar{\mathbf{y}} = \begin{bmatrix} \mathbf{A}\mathbf{s}(1) \\ \vdots \\ \mathbf{A}\mathbf{s}(T) \end{bmatrix}$ with $\mathbf{s}(t) = [s_1(t)\ s_2(t)]^T$ and $\mathbf{A} = \begin{bmatrix} \mathbf{a}(\omega_1) & \mathbf{b}(\omega_2, \phi) \end{bmatrix}$ where the signal sources $s_m(t)$ are viewed as deterministic known signals and the steering vectors are defined by $[\mathbf{a}(\omega_1)]_\ell = e^{i\omega_1 \ell}$ and $[\mathbf{b}(\omega_2, \phi)]_\ell = e^{i(\omega_2 \ell + \phi \ell^2)}$ with $\ell \in [0 : L-1]$. In addition, we assume $\omega_1 \neq \omega_2$. We define the determinist non-zero separation by $\delta = \omega_2 - \omega_1$. We focus our attention on the electric parameters but these parameters can be linked to the physical parameters, namely the DOA θ_m and the range r, according to $\omega_m = -2\pi d/\lambda \sin(\theta_m)$ and $\phi = \pi d^2/(\lambda r)\cos(\theta_2)^2$ where d is the distance inter-sensor and λ is the wavelength.

3. Analytic ARL in the Smith's sense

3.1. Analytical expression of the Cramer-Rao Bound

The Cramer-Rao Bound (CRB) verifies the covariance inequality principle [3]. This bound is largely used in the signal processing community since it gives the best performance in term of Mean Square Error (MSE) at high Signal to Noise Ratio (SNR). Let $\hat{\omega}_m$ be an unbiased estimate of ω_m, then

$$E\{(\omega_m - \hat{\omega}_m)^2\} \geq [\mathbf{C}]_{mm} \stackrel{\text{def}}{=} [\mathbf{J}^{-1}]_{mm} \tag{1}$$

where $E\{.\}$ is the mathematical expectation and $\mathbf{J}$ is the Fisher Information Matrix (FIM) defined by the Slepian-Bangs formula for a complex circular Gaussian observation $\mathbf{y} \sim \mathcal{CN}(\bar{\mathbf{y}}, \sigma^2 \mathbf{I})$:

$$[\mathbf{J}]_{ij} = \frac{2}{\sigma^2} \Re\left\{ \left(\frac{\partial \bar{\mathbf{y}}}{\partial \omega_i}\right)^H \frac{\partial \bar{\mathbf{y}}}{\partial \omega_j} \right\} = \frac{2}{\sigma^2} \sum_{t=1}^{T} \Re\left\{ \mathbf{s}(t)^H \left(\frac{\partial \mathbf{A}}{\partial \omega_i}\right)^H \frac{\partial \mathbf{A}}{\partial \omega_j} \mathbf{s}(t) \right\} \tag{2}$$

where $\frac{\partial \mathbf{A}}{\partial \omega_1} = \begin{bmatrix} \dot{\mathbf{a}}(\omega_1) & \mathbf{0} \end{bmatrix}$, $\frac{\partial \mathbf{A}}{\partial \omega_2} = \begin{bmatrix} \mathbf{0} & \dot{\mathbf{b}}(\omega_2, \phi) \end{bmatrix}$, $\frac{\partial \mathbf{A}}{\partial \phi} = \begin{bmatrix} \mathbf{0} & \ddot{\mathbf{b}}(\omega_2, \phi) \end{bmatrix}$ where $\dot{\mathbf{a}}(\omega_1)$ (*resp.* $\dot{\mathbf{b}}(\omega_2, \phi)$) is the first-order derivative of $\mathbf{a}(\omega_1)$ (*resp.* $\mathbf{b}(\omega_2, \phi)$) *w.r.t.* the parameters ω_1, ω_2, respectively and $\ddot{\mathbf{b}}(\omega_2, \phi)$ is the first-order derivative *w.r.t.* the parameter ϕ. After some algebra, one obtains the CRB for the parameters of interest and the coupling term:

$$\text{CRB}(\omega_1) = \frac{\sigma^2}{2} \frac{\beta}{\mathcal{Q}}, \tag{3}$$

$$\text{CRB}(\omega_2) = \frac{\sigma^2}{2} \frac{L_2 ||\mathbf{s}_1||^2 - \frac{\eta^2(\delta)}{L_3 ||\mathbf{s}_2||^2}}{\mathcal{Q}}, \tag{4}$$

$$\text{CRB}(\omega_1, \omega_2) = -\frac{\sigma^2}{2} \frac{\zeta(\delta) - \frac{\eta(\delta) L_3}{L_4}}{\mathcal{Q}}. \tag{5}$$

where $\beta = ||\mathbf{s}_2||^2 \left(L_2 - \frac{L_3^2}{L_4}\right)$, $\mathcal{Q} = \beta \left(L_2 ||\mathbf{s}_1||^2 - \frac{\eta^2(\delta)}{L_3 ||\mathbf{s}_2||^2}\right) - \left(\zeta(\delta) - \eta(\delta)\frac{L_2}{L_3}\right)^2$, $L_r = \sum_{\ell=0}^{L-1} \ell^r$ and $\eta(\delta) = \Re\left\{ h \sum_{\ell=0}^{L-1} \ell^2 e^{i(\delta \ell + \phi \ell^2)} \right\}$ and $\eta(\delta) = \Re\left\{ h \sum_{\ell=0}^{L-1} \ell^3 e^{i(\delta \ell + \phi \ell^2)} \right\}$ with $h = \mathbf{s}_1^H \mathbf{s}_2$.

3.2. Derivation of the analytic ARL

3.2.1. The Smith's equation: The Smith's methodology [7] provides the ARL as the solution of the following equation:

$$\text{CRB}(\delta) = \delta^2. \tag{6}$$

Expanding CRB(δ) *w.r.t.* the localization parameters and the coupling term, we have [8]:

$$\text{CRB}(\delta) = \text{CRB}(\omega_1) + \text{CRB}(\omega_2) - 2\text{CRB}(\omega_1, \omega_2) \tag{7}$$

where the CRBs are given by expressions (3)-(5).

3.2.2. Polynomial resolution of the linearized problem: Solving analytically equation (6) with (7) and (3)-(5) seems an intractable problem. Recalling that we assume that the separation δ is small then a first-order Taylor expansions of $\zeta(\delta)$ and $\eta(\delta)$ lead to the following approximations:

$$\eta(\delta) \approx \Re\left\{h(v(\phi) + j\delta r(\phi))\right\}, \quad \zeta(\delta) \approx \Re\left\{h(u(\phi) + j\delta v(\phi))\right\} \tag{8}$$

in which $u(\phi) = \sum_{\ell=0}^{L-1} \ell^2 e^{i\phi\ell^2}$, $v(\phi) = \sum_{\ell=0}^{L-1} \ell^3 e^{i\phi\ell^2}$ and $r(\phi) = \sum_{\ell=0}^{L-1} \ell^4 e^{i\phi\ell^2}$. The above relation can be rewritten in a linear form *w.r.t.* the separation according to

$$\Re\left\{h(v(\phi) + j\delta r(\phi))\right\} = P - \delta Q, \quad \Re\left\{h(u(\phi) + j\delta v(\phi))\right\} = P' - \delta Q' \tag{9}$$

where $P = \Re\{hv(\phi)\}$, $Q = -\Im\{hr(\phi)\}$, $P' = \Re\{hu(\phi)\}$ and $Q' = -\Im\{hv(\phi)\}$. Using these approximations, one can rewrite the CRB as

$$\mathbf{C} = \mathbf{J}^{-1} \stackrel{\text{def}}{=} \frac{\sigma^2}{2} \begin{bmatrix} \mathbf{Q}^{-1} & \times \\ \times & \times \end{bmatrix}, \tag{10}$$

where, the Schur complement can be approximated as follow $\mathbf{Q} \approx \begin{bmatrix} P_2(\delta) & P_1(\delta) \\ P_1(\delta) & \beta \end{bmatrix}$ in which we have introduced the two following polynomials $P_2(\delta) = \alpha_1\delta + \alpha_0$ and $P_2(\delta) = a_2\delta^2 + a_1\delta + a_0$ where $a_2 = -\frac{Q^2}{L_4||\mathbf{s}^2||^2}$, $a_1 = -\frac{2PQ}{L_4||\mathbf{s}^2||^2}$, $a_0 = L_2||\mathbf{s}_1||^2 - \frac{P^2}{L_4||\mathbf{s}^2||^2}$, $\alpha_1 = Q' - \frac{L_3}{L_4}Q$ and $\alpha_0 = P' - \frac{L_3}{L_4}P$. The linearized CRB expressions are now given by

$$\text{CRB}(\omega_1) \approx \frac{\sigma^2}{2} \frac{\beta}{Q(\delta)}, \tag{11}$$

$$\text{CRB}(\omega_2) \approx \frac{\sigma^2}{2} \frac{P_2(\delta)}{Q(\delta)}, \tag{12}$$

$$\text{CRB}(\omega_1, \omega_2) \approx -\frac{\sigma^2}{2} \frac{P_1(\delta)}{Q(\delta)}, \tag{13}$$

where $Q(\delta) = P_2(\alpha)\beta - P_1^2(\delta) = (\beta a_2 - \alpha_1^2)\delta^2 + (\beta a_1 - 2\alpha_0\alpha_1)\delta + \beta a_0 - \alpha_0^2$. Consequently from (6) and (11)-(13), the Smith's equation becomes $\delta^2 = \frac{Q'(\delta)}{Q(\delta)}$ where $Q'(\delta) = \frac{\sigma^2}{2}(\beta + P_2(\delta) + 2P_1(\delta)) \stackrel{\text{def}}{=} c_2\delta^2 + c_1\delta + c_0$ with $c_2 = \frac{\sigma^2}{2}a_2$, $c_1 = \frac{\sigma^2}{2}(a_1 + 2\alpha_1)$ and $c_0 = \frac{\sigma^2}{2}(\beta + a_0 + 2\alpha_0)$. So, it is easy to see that the Smith's equation provides the ARL as the solution of a 4-th order polynomial according to

$$R(x) = Q(x)x^2 - Q'(x) \stackrel{\text{def}}{=} x^4 + g_3x^3 + g_2x^2 + g_1x + g_0 \tag{14}$$

where x is a free variable with $g_0 = -\frac{c_0}{\beta a_2 - \alpha_1^2}$, $g_1 = -\frac{c_1}{\beta a_2 - \alpha_1^2}$, $g_2 = \frac{\beta a_0 - \alpha_0^2 - c_2}{\beta a_2 - \alpha_1^2}$, $g_3 = \frac{\beta a_1 - 2\alpha_0\alpha_1}{\beta a_2 - \alpha_1^2}$ where we assume $\beta a_2 \neq \alpha_1^2$.

3.2.3. Analytic solutions of $R(x)$: The resolution of a 4-th order polynomial is not straightforward but we propose a solution to this problem. More precisely, as $\mathrm{CRB}(\delta) = \mathrm{CRB}(-\delta)$, it is easy to see that if δ is a solution of the Smith's equation then $-\delta$ is also a solution. This implies that if δ is a root of $R(x)$ then $-\delta$ is also a root. So, $R(x)$ has four roots, namely $\{\delta, -\delta, r_1, r_2\}$ and a decomposition of this polynomial into a product of monomial terms is given by

$$\begin{aligned} R(x) &= (x-\delta)(x+\delta)(x-r_1)(x-r_2) && (15)\\ &= x^4 - (r_1+r_2)x^3 + (r_1r_2-\delta^2)x^2 && (16)\\ &+ \delta^2(r_1+r_2)x - r_1r_2\delta^2. && (17)\end{aligned}$$

We can identify the coefficients of the polynomials (14) and (17) according to

$$\begin{cases} -(r_1+r_2) &= \frac{\beta a_1 - 2\alpha_0\alpha_1}{\beta a_2 - \alpha_1^2},\\ r_1r_2 - \delta^2 &= \frac{\beta a_0 - \alpha_0^2 - c_2}{\beta a_2 - \alpha_1^2},\\ \delta^2(r_1+r_2) &= -\frac{c_1}{\beta a_2 - \alpha_1^2},\\ -r_1r_2\delta^2 &= -\frac{c_0}{\beta a_2 - \alpha_1^2}. \end{cases} \tag{18}$$

Combining the second and the last equations, we have to solve the following polynomial:

$$R'(x) = (\beta a_2 - \alpha_1^2)x^4 + (\beta a_0 - \alpha_0^2 - c_2)x^2 - c_0. \tag{19}$$

Polynomial $R'(x)$ can be reformulated as a 2-rd order polynomial according to

$$R'(z) = (\beta a_2 - \alpha_1^2)z^2 + (\beta a_0 - \alpha_0^2 - c_2)z - c0. \tag{20}$$

The discriminant is

$$\Delta = (\beta a_0 - \alpha_0^2 - c_2)^2\left(1 + 4\frac{(\beta a_2 - \alpha_1^2)c_0}{(\beta a_0 - \alpha_0^2 - c_2)^2}\right). \tag{21}$$

The study of the sign of the discriminant is not straightforward but observe that $c_0, c_2 \overset{\sigma^2\to 0}{\longrightarrow} 0$, so if the noise variance is low then the discriminant is given by

$$\Delta \overset{\sigma^2\to 0}{\longrightarrow} (\beta a_0 - \alpha_0^2 - c_2)^2 \geq 0. \tag{22}$$

So, we know that if the noise variance is not too high, there exists two candidates for the ARL (or a double solution if the discriminant is zero) which are given by

$$\delta = \sqrt{\frac{-(\beta a_0 - \alpha_0^2 - c_2) \pm \sqrt{\Delta}}{2(\beta a_2 - \alpha_1^2)}} \tag{23}$$

where $\sqrt{\Delta}$ is given by the square root of expression (21).

3.2.4. Analytic expression of the ARL: To discriminate the two possible solutions, we advocate that $\delta \overset{\sigma^2\to 0}{\longrightarrow} 0$ is a reasonable property. Thus, note thanks to (22), the sign in (23) must be chosen as positive to ensure $\delta \overset{\sigma^2\to 0}{\longrightarrow} 0$. If not (in case of negative sign in (23)) the chosen solution will be meaningless. So, finally the ARL is given by

$$\delta = \sqrt{\frac{-(\beta a_0 - \alpha_0^2 - c_2) + \sqrt{\Delta}}{2(\beta a_2 - \alpha_1^2)}}. \tag{24}$$

To provide further simplifications, we can see that

$$\sqrt{\Delta} \approx \beta a_0 - \alpha_0^2 - c_2 + \frac{2(\beta a_2 - \alpha_1^2)c_0}{\beta a_0 - \alpha_0^2 - c_2} \tag{25}$$

thanks to a first-order Taylor expansion of the square root $\sqrt{1+x} \approx 1 + 1/2x$ for small x. Using the above approximation of $\sqrt{\Delta}$, the ARL takes the simple following expression:

$$\delta \approx \sqrt{\frac{c_0}{\beta a_0 - \alpha_0^2 - c_2}}. \tag{26}$$

So, we can see that $\delta \approx O(\sigma)$.

4. Numerical illustrations

In this simulation part, we have considered an array constituted by $L = 10$ sensors. The sources are chosen to be close where $\theta_{FF} = \pi/3$ and $\theta_{NF} = \pi/3.1$ and the range of this source belongs to the interval $[0.62(d^3(L-1)^3/T)\lambda, 2d^2(L-1)^2/\lambda]$ [5] where $d = 0.0125$ m and the carrier frequency is $f_0 = 10$ MHz (the wavelength is $\lambda = c/f_0$ where c is celerity of the light). The modulus of the NF sources is fixed, for example, to ten times higher than the modulus of FF source (we can assume that the source which is the closest has a higher power than the one which belongs to the FF). The number of snapshot is $T = 100$. On Fig. 1-a, we have reported the positive roots of each polynomials $R(x)$, $R'(x)$ and the analytic ARL given in expression (24). As we can see one root for polynomials $R(x)$ and $R'(x)$ are independent from the noise variance. These two roots have to be ignored. In addition, we can see that the one root for $R(x)$ and $R'(x)$ are equal and follow a decreasing function *w.r.t.* the inverse of the noise variance. Moreover, the analytic ARL given in expression (24) is in good agreement with these roots and thus assess the validity of the derivations given in the previous section. The ARL given in expression (24) and the approximated ARL under the assumption of low noise variance given in expression (26) are reported on Fig. 1-b. It is important to highlight the good accuracy of the proposed approximated ARL.

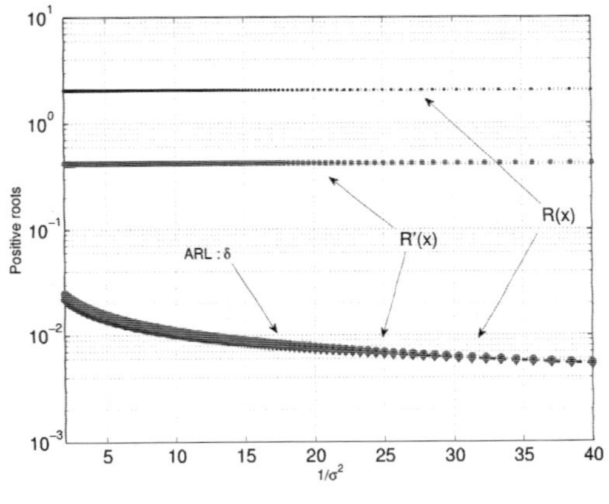

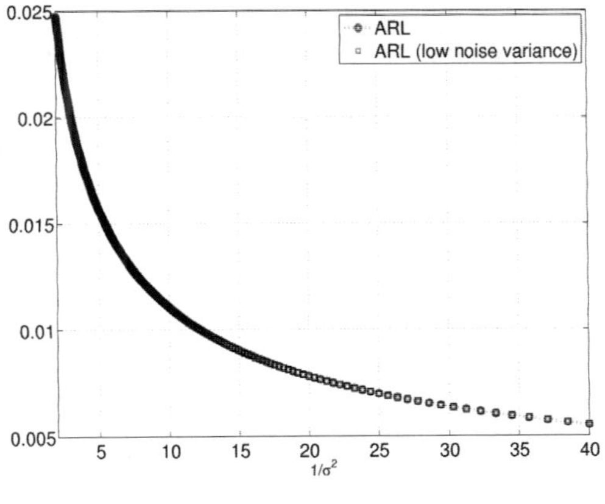

Figure 1. (a) Positive roots for polynomials $R(x)$, $R'(x)$ and analytic ARL given in expression (24) vs. the inverse of the noise variance. (b) ARL given in expression (24) and the approximated ARL under the assumption of low noise variance given in expression (26) vs. the inverse of the noise variance.

5. Conclusion

In this paper, we have derived and analyzed the Angular Resolution Limit (ARL) based on the resolution of the linearized Smith's equation for the new and realistic scenario where far-field and near-field sources are mixed. Generally speaking, the ARL is important and fundamental since this quantity gives the limit in the resolvability/separation of two closely spaced signals in term of their direction of arrivals. We show that for the chosen application, the resolution of the Smith's equation turns to be the selection of the "right" root of a 4-th order polynomial. This allows us to give a closed-form (analytic) expression of the ARL.

6. References

[1] H. Krim and M. Viberg, "Two decades of array signal processing research: the parametric approach", *IEEE Signal Processing Magazine*, Volume 13, Issue 4, July 1996.

[2] J. Liang and D. Liu, "Passive Localization of Mixed Near-Field and Far-Field Sources Using Two-stage MUSIC Algorithm", *IEEE Trans. Signal Process.*, Volume: 58 Issue: 1, Jan. 2010, pp: 108 - 120

[3] P. Stoica and R. L. Moses, *Spectral Analysis of Signals*, Prentice Hall, 2005.

[4] Y. D. Huang and M. Barkat, "Near-field multiple source localization by passive sensor array", *IEEE Trans. Antennas Propag.*, vol. 39, no. 7, pp. 968-975, Jul. 1991.

[5] E. Grosicki, K. Abed-Meraim, and Y. Hua, "A weighted linear prediction method for near-field source localization", *IEEE Trans. Signal Process.*, vol. 53, pp. 3651- 3660, Oct. 2005.

[6] M. N. El Korso, G. Bouleux, R. Boyer, and S. Marcos, "Sequential estimation of the range and the bearing using the zero-forcing MUSIC approach", in *Proc. EUSIPCO*, Glasgow, Scotland, Aug. 2009, pp. 1404-1408.

[7] S. T. Smith, "Statistical resolution limits and the complexified Cramer-Rao bound", *IEEE Trans. Signal Process.*, vol. 53, pp. 1597-1609, May 2005.

[8] M. N. El Korso, R. Boyer, A. Renaux and S. Marcos, "Statistical resolution limit for the multidimensional harmonic retrieval model: Hypothesis test and Cramer-Rao bound approaches", *EURASIP Journal on Advances in Signal Processing, special issue "Advances in Angle-of-Arrival and Multidimensional Signal Processing for Localization and Communications"*, no. 12, pp. 1-14, doi:10.1186/1687 − 6180 − 2011 − 12, 2011.

[9] R. Boyer, "Performance bounds and angular resolution limit for the moving co-located MIMO radar", *IEEE Trans. Signal Process.*, vol. 59, no. 4, pp. 1539-1552, 2011.

[10] M. N. El Korso, R. Boyer, A. Renaux and S. Marcos, "Statistical resolution limit of the uniform linear cocentered orthogonal loop and dipole array", *IEEE Trans. Signal Process.*, Volume: 59, Issue: 1, Jan. 2011, pp. 425-431.

[11] W. Zhi and M. Chia, "Near-field source localization via symmetric subarrays", *IEEE Signal Process. Lett.*, vol. 14, no. 6, pp. 409-412, 2007.

[12] M. N. El Korso, R. Boyer, A. Renaux and S. Marcos, "Conditional and Unconditional Cramér Rao Bounds for Near-Field Source Localization", *IEEE Trans. Signal Process.*, Volume: 58, Issue: 5, May. 2010, pp. 2901-2907.

Journal of Physics: Conference Series **464** (2013) 012003 doi:10.1088/1742-6596/464/1/012003

Regularized Blind Deconvolution with Poisson Data

Loïc Lecharlier[1] **and Christine De Mol**[2]

[1] Unité SIMa, Université de Liège Gembloux Agro-Bio Tech, Passage des Déportés 2, 5030 Gembloux, Belgium

[2] Department of Mathematics and ECARES, Université Libre de Bruxelles, Campus Plaine CPI 217, Boulevard du Triomphe, 1050 Brussels, Belgium

E-mail: loic.lecharlier@ulg.ac.be, demol@ulb.ac.be

Abstract. We propose easy-to-implement algorithms to perform blind deconvolution of nonnegative images in the presence of noise of Poisson type. Alternate minimization of a regularized Kullback-Leibler cost function is achieved via multiplicative update rules. The scheme allows to prove convergence of the iterates to a stationary point of the cost function. Numerical examples are reported to demonstrate the feasibility of the proposed method.

1. Introduction

Alternating multiplicative update rules have been popularized recently by Lee and Seung [1] for the task of nonnegative matrix factorization, i.e. given Y a $n \times m$ nonnegative matrix of observations, the task of finding a $n \times p$ nonnegative matrix K and a $p \times m$ nonnegative matrix X such that $Y = KX$. These rules can be derived by minimizing either a least-squares criterion or the following generalized Kullback-Leibler divergence between Y and KX:

$$KL(Y, KX) = \sum_{i=0}^{m-1} \sum_{j=0}^{n-1} \left[Y_{i,j} \ln \left(\frac{Y_{i,j}}{(KX)_{i,j}} \right) - Y_{i,j} + (KX)_{i,j} \right].$$

The minimization of this latter cost function is the usual model for the case where the data Y are affected by photon-counting noise, i.e. obey a Poisson distribution. The minimization is performed alternately with respect to K and X with the following update rules, for $k = 0, 1, \ldots$, starting from arbitrary but strictly positive matrices $K^{(0)}$ and $X^{(0)}$:

$$K^{(k+1)} = \frac{K^{(k)}}{\mathbf{1}_{n\times m}(X^{(k)})^T} \circ \frac{Y}{K^{(k)}X^{(k)}}(X^{(k)})^T; \; X^{(k+1)} = \frac{X^{(k)}}{(K^{(k+1)})^T \mathbf{1}_{n\times m}} \circ (K^{(k+1)})^T \frac{Y}{K^{(k+1)}X^{(k)}},$$

where $\mathbf{1}_{n\times m}$ is a $n \times m$ matrix of ones, X^T is the transpose of X and we use the Hadamard (entrywise) product $\circ$ and division. Monotonic decrease of the cost function is guaranteed by the fact that the algorithm can be derived as a Majorization-Minimization (MM) scheme, through the use of surrogate cost functions [1].

In the present paper, we focus on a special instance of this problem namely blind deconvolution in incoherent optical imaging, where the original image X, the blurred image Y and the space-invariant point spread function (PSF) K are nonnegative light intensities. In this case the matrix multiplications are to be replaced by circular convolutions (assuming

$m = p = n$). If the PSF K is known, the above algorithm is well known in astronomy as Richardson-Lucy's one [2, 3] and in medical imaging as (EM)ML (Expectation-Maximization Maximum Likelihood). For the special case of blind deconvolution, the above alternate minimization algorithm has been discussed previously in the literature by different authors dating back to [4, 5].

To prevent the algorithm to converge to the trivial solution where the PSF is a Dirac delta peak and the image is the original blurred one, we consider regularized versions of the alternating Richardson-Lucy algorithm, adding appropriate penalty terms to the Kullback-Leibler divergence. In addition we impose a flux normalization condition on the PSF. We can then establish the convergence of the iterates to a stationary point of the corresponding cost function which is separately but not jointly convex with respect to the image and the PSF. Note that for this problem convergence results are rare in the literature. Recent exceptions are the blind imaging methods based on the SGP algorithm of [6] and proposed in [7] for space-variant blind image restoration and in [8] for post-adaptive-optics imaging in astronomy, as well as the inexact block coordinate descent method of [9]. These algorithms and convergence results, however, are not exactly of the same kind as ours. The alternate proximal algorithm of [10] applies to the case of Gaussian noise (least-squares discrepancy) but could also be applied to Poisson noise when a quadratic approximation of the Kullback-Leibler divergence is used.

2. Blind deconvolution algorithms

Let X be the $n \times n$ matrix formed by the 2D image intensities $X_{i,j}$ in pixel (i, j) and K the $n \times n$ matrix containing the values of the space-invariant PSF. By $K * X$ we denote their 2D circular convolution which can be computed by going to Fourier space by means of a FFT. Besides $\tilde{X}_{i,j} \equiv X_{n-i,n-j}$. Assuming $Y_{i,j} \geq 0, \forall i, j$, we want to solve the following minimization problem:

$$\text{find} \quad \operatorname*{argmin}_{K,X} F(K, X) = KL(Y, K * X) + \frac{\mu}{2}\|K\|_F^2 + \lambda \sum_{i=0}^{n-1}\sum_{j=0}^{n-1} X_{i,j} + \frac{\nu}{2}\|X\|_F^2$$

under the constraints $X_{i,j} \geq 0$, $K_{i,j} \geq 0$ and $\sum_{i=0}^{n-1}\sum_{j=0}^{n-1} K_{i,j} = 1$, with nonnegative regularization parameters λ, μ, ν. Notice that we penalize the squared Frobenius norm $\|K\|_F^2 = \sum_{i=0}^{n-1}\sum_{j=0}^{n-1} K_{i,j}^2$ of the PSF and both the Frobenius and the sparsity-enforcing ℓ^1-norm of the image. This allows to cover the cases of pure quadratic (Tikhonov) regularization when $\lambda = 0$, of the so-called "lasso" regularization when $\nu = 0$ as well as of a combination of both penalties which goes under the name "elastic-net" after [11]. As we noticed in practice, the latter strategy seems to perform better in the recovery of neighboring pixel values which are correlated.

To solve the optimization problem, we use the following alternating minimization algorithm which can be easily derived using the usual surrogate for the Kullback-Leibler divergence and the fact that the penalties are separable, i.e. are a sum of terms, each of which depends only on a single unknown scalar component:

1) initialize $X^{(0)}$ and $K^{(0)}$ with arbitrary strictly positive values, satisfying the normalization constraint $\displaystyle\sum_{i=0}^{n-1}\sum_{j=0}^{n-1} K_{i,j}^{(0)} = 1$;
2) at each iteration
   i) introduce a Lagrange parameter α for the normalization constraint and, for $\mu \neq 0$, determine its value by solving the following equation

$$2\mu + n^2\left(\alpha + \sum_{l=0}^{n-1}\sum_{m=0}^{n-1} X_{l,m}^{(k)}\right) - \sum_{i=0}^{n-1}\sum_{j=0}^{n-1}\sqrt{\left(\alpha + \sum_{l=0}^{n-1}\sum_{m=0}^{n-1} X_{l,m}^{(k)}\right)^2 + 4\mu K_{i,j}^{(k)}\left(\frac{Y}{K^{(k)} * X^{(k)}} * \tilde{X}^{(k)}\right)_{i,j}} = 0$$

by means of a few Newton-Raphson iterations, whereas for $\mu = 0$,

use $\alpha = \sum_{l=0}^{n-1}\sum_{m=0}^{n-1} Y_{l,m} - \sum_{l=0}^{n-1}\sum_{m=0}^{n-1} K_{l,m}^{(k)}$,

ii) compute $A^{(k)} = K^{(k)} \circ \left(\frac{Y}{K^{(k)} * X^{(k)}} * \tilde{X}^{(k)} \right)$,

iii) compute $B^{(k)} = \mathbf{1}_{n\times n} X^{(k)} \mathbf{1}_{n\times n} + \alpha \mathbf{1}_{n\times n}$,

iv) set $K^{(k+1)} = \frac{2A^{(k)}}{B^{(k)} + \sqrt{B^{(k)} \circ B^{(k)} + 4\mu A^{(k)}}}$,

v) compute $C^{(k+1)} = X^{(k)} \circ \left(\tilde{K}^{(k+1)} * \frac{Y}{K^{(k+1)} * X^{(k)}} \right)$,

vi) compute $D^{(k+1)} = \mathbf{1}_{n\times n} K^{(k+1)} \mathbf{1}_{n\times n} + \lambda \mathbf{1}_{n\times n}$,

vii) set $X^{(k+1)} = \frac{2C^{(k+1)}}{D^{(k+1)} + \sqrt{D^{(k+1)} \circ D^{(k+1)} + 4\nu C^{(k+1)}}}$.

Notice that the nonnegativity of both the image and the PSF is automatically preserved throughout the multiplicative update scheme.

Building upon Zangwill's convergence theory [12], we can establish the following properties for the above algorithm.

Theorem
If Y is a nonnegative matrix and $X^{(0)}$, $K^{(0)}$ are two strictly positive matrices, with normalization $\sum_{i=0}^{n-1}\sum_{j=0}^{n-1} K_{i,j}^{(0)} = 1$, then

a) $F\left(K^{(k+1)}, X^{(k+1)}\right) \leq F\left(K^{(k)}, X^{(k)}\right)$;

b) $F\left(K^{(k+1)}, X^{(k+1)}\right) < F\left(K^{(k)}, X^{(k)}\right)$ *if and only if* $\left(K^{(k+1)}, X^{(k+1)}\right) \neq \left(K^{(k)}, X^{(k)}\right)$;

c) *the sequence of iterates* $\left(K^{(k)}, X^{(k)}\right)$ *converges to a stationary point* (K^*, X^*) *of* $F(K, X)$.

The detailed proof of this convergence theorem is too long to be reported here. Hence we just sketch the key elements to be established:

(i) prove that all the fixed points of the algorithm are stationary points i.e. verify the first order Karush-Kuhn-Tucker conditions;

(ii) prove that all the iterates lie in a compact set;

(iii) using points a) and b) above, Zangwill's theorem allows to conclude that all the accumulation points of the sequence of iterates are stationary points;

(iv) finally, show that the sequence cannot oscillate between different stationary points.

Our results can be extended to cover the case of the following smoothed total-variation (TV) penalty on the image

$$TV(X) = \sum_{i=0}^{n-1}\sum_{j=0}^{n-1} \left|\nabla_{i,j} X\right|_{\varepsilon} \text{ with } \left|\nabla_{i,j} X\right|_{\varepsilon} = \sqrt{\varepsilon^2 + (X_{i+1,j} - X_{i,j})^2 + (X_{i,j+1} - X_{i,j})^2}$$

using the separable surrogate for this penalty introduced in [13]. Moreover, we can derive similar algorithms and convergence results for the case of a Gaussian noise, i.e. when the Kullback-Leibler divergence is replaced by a quadratic Frobenius norm. To the best of our knowledge, also in the non-blind case, the algorithm with a smoothed TV penalty is new for the case of Poisson noise whereas our convergence results are original for both cases of Gaussian and Poisson noise.

3. Numerical simulations

We have investigated the numerical performance of our method by testing it on various image restoration problems and we report a few examples here. As test images, we use the spiral nebula NGC 7027 (256×256), the satellite (256×256) and the cameraman (512×512). The blurred image is generated by circular low-pass filtering, i.e. by convolution with an Airy pattern, and corrupted by Poisson noise which results in a root-mean-square error (RMSE) of 1% for the nebula and the satellite and of 2.5 % for the cameraman.

Figure 1. The original images and the original PSF

Figure 2. The blurred and noisy images, the reconstructed images and the reconstructed PSFs

For the satellite and the cameraman, we used a TV penalty on the image and for the nebula an elastic-net penalty. The algorithm was initialized with the blurred and noisy image and with a uniform PSF. The original images and PSF are reported in Figure 1, the reconstructions in Figure 2 whereas the values of all parameters used are given in Table 1, together with the number of iterations performed, the computing time in MATLAB on a laptop and the relative RMS reconstruction errors on both the image X and the PSF K. For a better visualization of the PSF, only the central part of the image (zooming by a factor 4 in both dimensions) is displayed in false colors.

Image	μ	λ	ε / ν	It	Time	$RMSE_K$	$RMSE_X$
Nebula	10^9	10^{-7}	$6 \cdot 10^{-8}$	2000	12m37s	0.30	0.33
Satellite	$1.5 \cdot 10^6$	0.0485	$\sqrt{10^{-7}}$	200	1m46s	0.20	0.18
Cameraman	10^7	0.03	$\sqrt{10}$	2000	54m30s	0.17	0.15

Table 1. Parameters used for the simulations

Acknowledgments

This work was supported by the research grants ARC-AUWB/2010-15/ULB-11, GOA-VUB-062 and IAP P7/06 StUDys.

References

[1] Lee D D and Seung H S 2000 *Advances in Neural Information Processing Systems* **13** 556–562
[2] Richardson W H 1972 *J. Opt. Soc. Am.* **62**(1) 55–59
[3] Lucy L B 1974 *The Astronomical Journal* **79** 745–754
[4] Holmes T J 1992 *J. Opt. Soc. Am.* **A9** 1052–1061
[5] Fish D A, Brinicombe A M, Pike E R and Walker J G 1995 *J. Opt. Soc. Am.* **A12** 58–65
[6] Bonettini S, Zanella R and Zanni L 2009 *Inverse Problems* **25**(1) 015002
[7] Ben Hadj S, Blanc-Féraud L and Aubert G 2012 Space variant blind image restoration Tech. Rep. RR-8073 INRIA
[8] Prato M, La Camera A, Bonettini S and Bertero M 2013 *Inverse Problems* **29** (in press)
[9] Bonettini S 2011 *IMA Journal of Numerical Analysis* **31**(4) 1431–1452
[10] Bolte J, Combettes P and Pesquet J C 2010 in *2010 IEEE International Conference on Image Processing (ICIP 2010)* pp 1673–1676
[11] Zou H and Hastie T 2005 *Journal of the Royal Statistical Society, Series B* **67** 301–320
[12] Zangwill W I 1969 *Nonlinear Programming: A Unified Approach* 1st ed (Prentice-Hall)
[13] Defrise M, Vanhove C and Liu X 2011 *Inverse Problems* **27**(6) 065002

Image restoration using sparse approximations of spatially varying blur operators in the wavelet domain

Paul Escande[1], **Pierre Weiss**[1], **and François Malgouyres**[2]

[1] ITAV-USR3505, Université de Toulouse, CNRS, Toulouse, France.
[2] IMT-UMR5219, Université de Toulouse, CNRS, Toulouse, France.

E-mail: paul.escande@gmail.com, pierre.armand.weiss@gmail.com, francois.malgouyres@math.univ-toulouse.fr

Abstract. Restoration of images degraded by spatially varying blurs is an issue of increasing importance in the context of photography, satellite or microscopy imaging. One of the main difficulty to solve this problem comes from the huge dimensions of the blur matrix. It prevents the use of naive approaches for performing matrix-vector multiplications. In this paper, we propose to approximate the blur operator by a matrix sparse in the wavelet domain. We justify this approach from a mathematical point of view and investigate the approximation quality numerically. We finish by showing that the sparsity pattern of the matrix can be pre-defined, which is central in tasks such as blind deconvolution.

1. Introduction

The problem of image restoration in the presence of spatially varying blurs drew a considerable attention in the context of satellite imaging with the Hubble space telescope [1]. It is now becoming increasingly important in various imaging modalities such as 3D fluorescence microscopy [2]. In this paper, we aim at providing new computational tools to tackle this problem. We consider a blurring operator H in $\Omega \subset \mathbb{R}^d$, defined for any $u \in \mathbb{L}^2(\Omega)$ as the following integral operator,

$$\forall x \in \Omega, \quad Hu(x) = \int_{y\in\Omega} K(x,y)u(y)dy, \tag{1}$$

where $\Omega \subset \mathbb{R}^d$ is the image domain. The function K is a spatially varying kernel defining the Point Spread Function (PSF) at each location $x \in \Omega$.

The naive computation of a product Hu is simply obtained by direct discretization of (1) and costs $O\left(N^{2d}\right)$ arithmetic operations for an image of dimension d with N pixels in each dimension. It is unsuitable for large scale problems encountered in imaging. Previous works attempted to reduce this computational cost by approximating H using either tensor products or piecewise convolutions [3, 1, 4]. Unfortunately, these strategies do not have a property that is highly desirable in numerical analysis: the ability to approximate the original operator with an arbitrary precision. In practice, we observed that they are often too coarse for complex situations encountered in microscopy imaging.

In a recent work [5], we proposed to approximate H by an operator diagonal in the wavelet domain. However, this approximation was shown to be too coarse for some practical applications. Our main contribution in this paper consists of showing that the diagonal approximation can be replaced by a *sparse* approximation and lead to arbitrarily close approximations. More precisely, we show that $\|H - \Psi S \Psi^*\|$, can be made arbitrarily small if Ψ^* denotes a forward wavelet transform and S is a sparse matrix that contains only $O(N^d)$ non zero coefficients instead of N^{2d}. As a practical consequence, matrix-vector products using the proposed approach costs $O\left(N^d\right)$ arithmetic operations, which is doable even for very large images. We also show that the sparsity pattern of S can be known a priori, which is a highly desirable property in the case of blind deconvolution, where the operator should be inferred from the blurred images.

In section 2, we recall the theoretical results motivating this approximation. Section 3.1, contains preliminary restoration results in order to illustrate theoretical results and to identify a number of coefficients offering a compromise between the storage costs and the restoration quality. In this section, the kept coefficients are obtained by simple thresholding and can have arbitrary coordinates. In section 3.2, we finally show that the sparsity pattern of the matrix Θ can be pre-defined and that this approach leads to almost equivalent results to a thresholding.

2. Theoretical motivations

Sparse approximations of integral operators have been theoretically analyzed in [6, 7]. Surprisingly this approach was never applied to the approximation of spatially varying blur operators. The closest application found in the literature deals with foveation [8], but it is not adapted to a large class of kernels.

We define an orthonormal wavelet basis of $\mathbb{L}^2(\mathbb{R})$ as the family of functions:

$$\left\{\psi_{j,m}\right\}_{j,m\in\mathbb{Z}}.$$

Each $\psi_{j,m}$ is a dilated and translated of the mother wavelet ψ,

$$\psi_{j,m}(t) = \sqrt{2^{-j}}\psi\left(2^{-j}t - m\right).$$

Let us recall a typical result that motivates the proposed approach. We stick to the one-dimensional case for the ease of exposition. Since H is a linear operator in a Hilbert space, it can be written as $H = \Psi\Theta\Psi^*$, where $\Theta : l^2 \to l^2$ is the matrix representation of the blur operator in the wavelet domain. Going from an infinite dimensional setting to a finite dimensional one can be performed by using projectors on linear subspaces of the multi-resolution analysis. Matrix Θ is characterized by the coefficients:

$$\theta_{j,m,k,n} := \left(\langle H\psi_{j,m}, \psi_{k,n}\rangle\right), \qquad \forall j,m,k,n \in \mathbb{Z}.$$

Theorem 1 (Decay of $\theta_{j,m,k,n}$ – [6]). *Supposing that:*

- *the wavelets are compactly supported with $supp(\psi_{j,m}) = I_{j,m}$,*
- *the wavelets have M vanishing moments,*
- *the operator H belongs to the class of Calderon-Zygmund operators, meaning that K satisfies:*

$$\exists C_M > 0, \text{ such that }, \forall x,y \in \Omega, \quad \begin{cases} |K(x,y)| \leq \dfrac{1}{|x-y|}, \\ \left|\partial_x^M K(x,y)\right| + \left|\partial_y^M K(x,y)\right| \leq \dfrac{C_M}{|x-y|^{1+M}}, \end{cases} \tag{2}$$

then the coefficients $\theta_{j,m,k,n}$ satisfy the following estimate:

$$\left|\theta_{j,m,k,n}\right| \leq C_M 2^{-\frac{|j-k|}{2}} \left(\frac{2^{\min(j,k)}}{dist(I_{j,m}, I_{k,n})}\right)^{M+1}, \tag{3}$$

with C_M a constant depending on M and the wavelets, $dist(I_{j,m}, I_{k,n})$ denotes the distance between the two supports of the wavelets $\psi_{j,m}$ and $\psi_{k,n}$.

Moreover, if the kernel K is compactly supported, then for sufficiently large $|m-n|$,

$$\theta_{j,m,k,n} = 0. \tag{4}$$

This results ensures that the coefficients away from the diagonal fastly decay to zero. A more precise analysis shows that Θ contains only $O\left(N\varepsilon^{-1/(M+1)}\right)$ above ε [6, 9]. Figure 1 shows the matrix Θ associated to a typical spatially varying blur operator in log scale. It is readily seen that the decrease of the coefficients away from the diagonal is extremely fast.

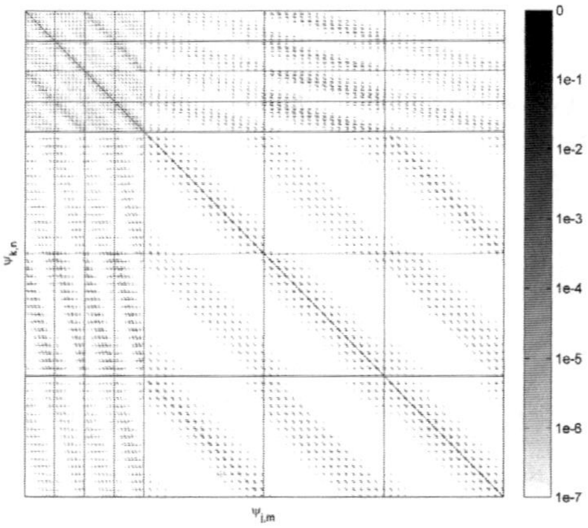

Figure 1. An example of a Θ matrix in a $\log_{10}$ scale for a 256×256 images. The matrix contains about 4.3 billion coefficients.

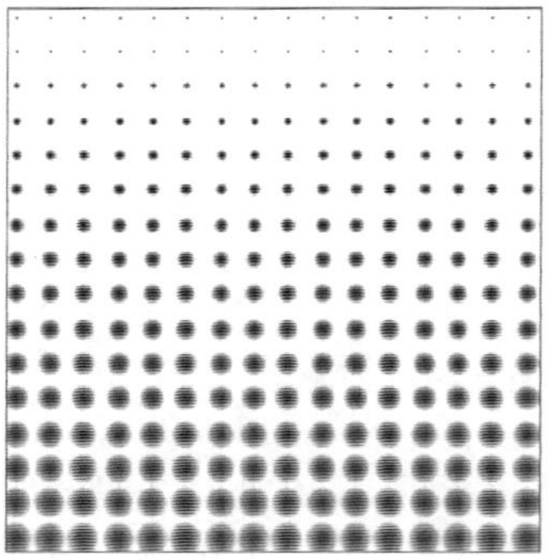

Figure 2. The point spread function of the blur used to generate the matrix on the left. The kernel is a Gaussian with equal variances increasing in the vertical direction.

In the next section, we reconstruct degraded images from the knowledge of various thresholded Θ matrices.

3. Restoration results

3.1. Restoration using thresholded matrices

In a first experiment, the Θ matrix is thresholded in order to keep a number of coefficients being a multiple of N^d (the number of pixels of the image). That is, we define the thresholded matrix Θ_{T_k} by zeroing all but the $T_k = kN^d$ largest coefficients of Θ. This experiment allows the identification of an admissible number of coefficients offering a proper compromise between the storage costs and the image restoration quality. To achieve this experiment, a full Θ matrix is numerically computed and then successively thresholded for various $k \in \mathbb{N}^*$. Thereafter, the image is restored assuming the following classical degradation model:

$$v = Hu + \eta, \quad \eta \sim \mathcal{N}(0, \sigma^2 I_{n^d}),$$

where v is the degraded image observed (see Figure 3c), u is the image to restore (see Figure 3a), and H is a blurring operator corresponding to the PSF displayed in Figure 3d. A standard TV-L2 optimization problem is solved to restore the image u:

$$\text{Find } u^* \in \underset{u \in \mathbb{R}^{n^d}, \|\Psi\Theta_{T_k}\Psi^* u - v\|_2^2 \leq \sigma^2 n}{\arg\min} TV(u), \tag{5}$$

where $TV(u)$ is the isotropic total variation of u.

Figure 4 displays restored images for various values of k and the sparsity pattern of Θ_{T_k}. As expected, considering more coefficients improves the deblurring results. However, in the presence of noise, the two images restored using the full matrix and the thresholded matrix with $k = 20$, provide very similar results. Therefore, it is possible to store only a small amount of coefficients and obtain results close to the ones obtained having a full (but unrealistic) knowledge of the blur matrix. Furthermore, these results are obtained drastically faster, see the speed-up Figure 4.

Journal of Physics: Conference Series **464** (2013) 012004 doi:10.1088/1742-6596/464/1/012004

Remark: it is near impossible to solve the deblurring problem with an exact operator due to the numerical burden of computing matrix-vector products in large dimensions. We thus chose a spatially varying blur that is sparse in the wavelet domain in order to be able to make comparisons between an exact computation and an approximated one. Speed up times might thus be much higher for other kinds of blurs.

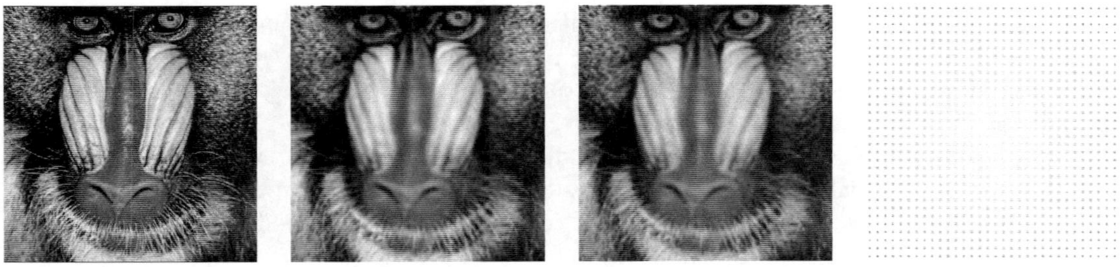

(a) Original Image (b) Blurred Image (c) Degraded Image $\sigma = 2.10^{-2}$, $SNR = 16.54$dB. (d) PSF

Figure 3. Images involved in the deblurring problem

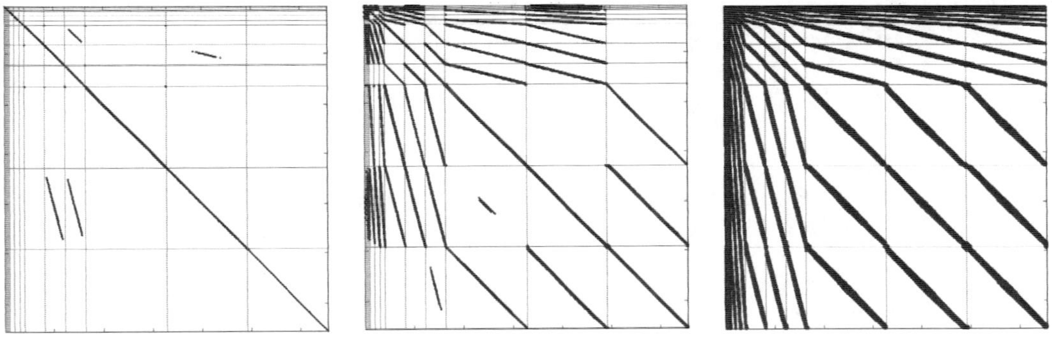

(a) Non-zeros coefficient for $k = 1$ (b) Non-zeros coefficient for $k = 20$ (c) Full Θ matrix

(d) Deblurred image for $k = 1$. $SNR = 17.12$dB, Speed-up = 22. (e) Deblurred image for $k = 20$. $SNR = 18.69$dB, Speed-up = 17. (f) Deblurred image with the full Θ, $k \simeq 1550$, $SNR = 18.93$dB.

Figure 4. Deblurred images using thresholded matrices

3.2. Restoration using pre-defined sparsity patterns

We are currently developping strategies to solve blind deconvolution problems using the proposed approximation. In this setting, it is not possible to threshold the exact (unknown) matrix to obtain a sparsity pattern. The knowledge of such a structure is primordial to approximate K using prior information. Our aim in this section is to show that sparsity patterns can be pre-defined, allowing convincing restoration results.

Our main observation is that for a given wavelet $\psi_{j,m}$, only its neighbours in the same scale, the finer scale and the coarser one lead to significant correlation coefficients $\langle H\psi_{j,m}, \psi_{k,n}\rangle$. This concept of neighbours is also supported by Theorem 1. Thus, sparsity patterns can be defined using notions of intra and inter-scale neighbourhoods.

For each node of the wavelet decomposition quadtree, a neighbourhood is defined. It describes which correlation coefficients shall be preserved to generate the sparsity pattern. Figure 5 illustrates this idea. In this example, we consider a wavelet transform of depth 2. A neighbourhood $\mathcal{N}_i$ is associated to each sub-band. For instance $\mathcal{N}_1$ can be represented as follows:

$$\mathcal{N}_1 = \begin{pmatrix} 2 & 2 & 2 & 2 & 2 & 2 & 2 & 2 \\ l & l & l & l & l & h & v & d \\ 0 & -1 & 1 & 0 & 0 & 0 & 0 & 0 \\ 0 & 0 & 0 & -1 & 1 & 0 & 0 & 0 \end{pmatrix} \left| \begin{array}{l} \text{scale} \\ \text{oriention} \\ \text{vertical translation} \\ \text{horizontal translation} \end{array} \right.$$

where l stands for the low frequency wavelet, h, v, d for the horizontal, vertical and diagonal orientations respectively. Figure 6 illustrates this neighbourhood $\mathcal{N}_1$ for a given wavelet at the center of the image.

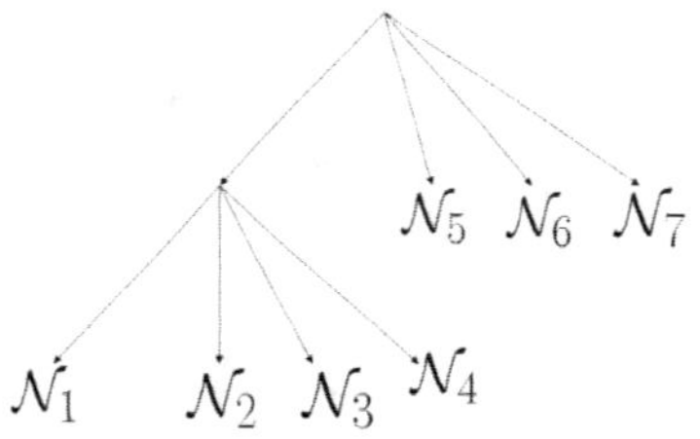

Figure 5. Quadtree corresponding to a wavelet transform of depth 2

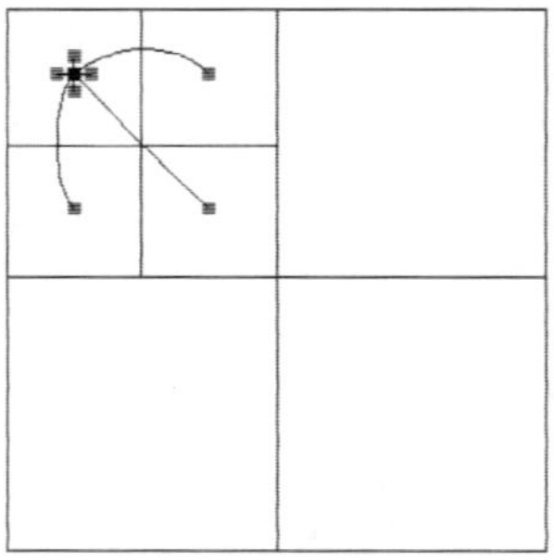

Figure 6. Illustration of the neighbourhood $\mathcal{N}_1$. In black, the given wavelet at the center of the image, and in red all the preserved neighbouring wavelets.

In order to validate this notion of neighbourhood, we considered two different scenarii. The first one consists of setting for all i, $\mathcal{N}_i = \begin{pmatrix} \text{same scale} \\ \text{all orientations} \\ 0 \\ 0 \end{pmatrix}$. The second one consists of setting for all i,

$$\mathcal{N}_i = \begin{pmatrix} \text{same scale} & \text{same scale} & \text{same scale} & \text{same scale} & \text{same scale} \\ \text{all orientations} & \text{all orientations} & \text{all orientations} & \text{all orientations} & \text{all orientations} \\ 0 & -1 & 1 & 0 & 0 \\ 0 & 0 & 0 & -1 & 1 \end{pmatrix},$$

Figure 7 displays restored images using sparsity patterns generated from these two neighbourhoods. This experiment shows that sparsity patterns defined from the correlations between neighbouring wavelets are relevant for our aim: approximating and interpolating the blurring operator.

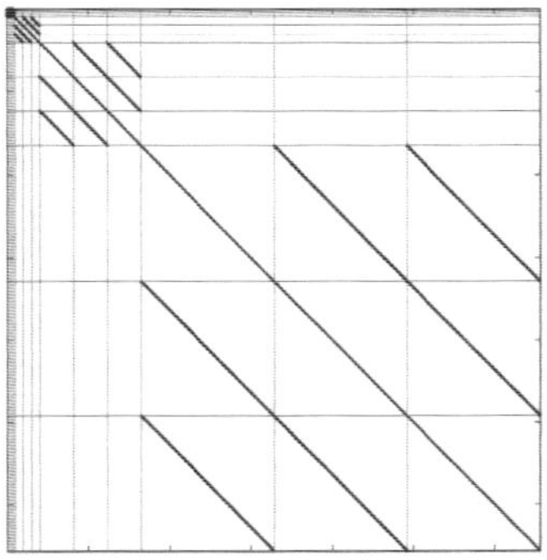

(a) Non-zero coefficients of pattern 1

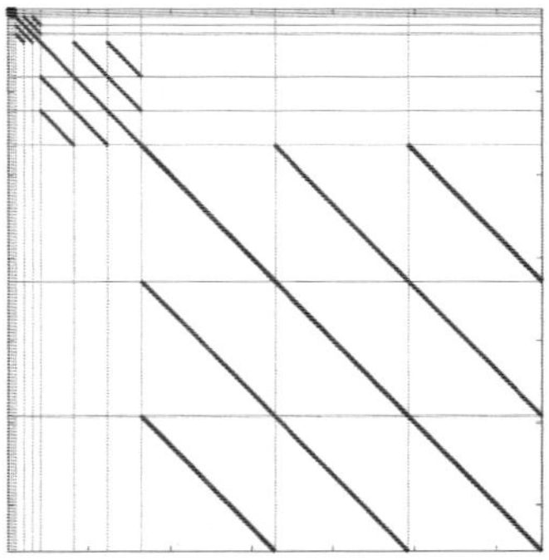

(b) Non-zero coefficients of pattern 2

(c) Deblurred image with pattern 1, $k \simeq 3$, $SNR = 16.74$dB.

(d) Deblurred image with pattern 2, $k \simeq 15$, $SNR = 18.30$dB.

Figure 7. Deblurred images using matrices defined by sparsity patterns

Conclusion

In this paper we investigated the use of sparse matrices in the wavelet domain in order to approximate spatially varying blur operators. We observed that very large compression factors still allow to obtain near optimal reconstruction results. Compared to previously proposed approach, this technique allows to approximate the exact operator with an arbitrary precision. It is thus suited to noiseless and noisy problems. We also showed that pre-established sparsity patterns can be used to efficiently approximate the blurring integral operator. This property is central in order to tackle blind deconvolution problems. We are currently working on this topic.

Acknowledgments

This project was funded by ANR SPHIM3D.

References

[1] Nagy J and O'Leary D 1998 *SIAM Journal on Scientific Computing* **19** 1063
[2] Ben Hadj S and Blanc-Féraud L 2011 URL http://hal.inria.fr/inria-00602650/fr/
[3] Zhang B, Zerubia J and Olivo-Marin J 2007 *Applied Optics* **46** 1819–1829
[4] Hansen P, Nagy J and O'leary D 2006 *Deblurring images: matrices, spectra, and filtering* (Siam)
[5] Escande P, Weiss P and Malgouyres F 2013 *Proceedings of ICPRAM* URL http://hal.archives-ouvertes.fr/hal-00733194
[6] Beylkin G, Coifman R and Rokhlin V 1991 *Commun. Pure and Applied Math.* **44** 141–183
[7] Coifman R and Meyer Y 1997 *Cambridge Studies in Advanced Math* **48**
[8] Chang E C, Mallat S and Yap C 1999 *Applied and Computational Harmonic Analysis* **9** 312–335
[9] Meyer Y and Salinger D H 1992 *Wavelets and operators* vol 37 (Cambridge Univ Press)

Journal of Physics: Conference Series **464** (2013) 012005 doi:10.1088/1742-6596/464/1/012005

Iterative choice of the optimal regularization parameter in TV image deconvolution

B Sixou[1], A Toma[1], L Denis[2] and F Peyrin[1,3]

[1] CREATIS, INSA-Lyon; Inserm, U1044; CNRS UMR 5220; Université Lyon 1; Université de Lyon; 69621, Lyon, France

[2] LaHC, CNRS UMR 5516, Université de Saint-Etienne, 42000 Saint-Etienne, France

[3] ESRF, Imaging Group, 38043 Grenoble Cedex, France

E-mail: bruno.sixou@insa-lyon.fr, alina.toma@creatis.insa-lyon.fr, loic.denis@univ-st-etienne.fr, peyrin@esrf.fr

Abstract. We present an iterative method for choosing the optimal regularization parameter for the linear inverse problem of Total Variation image deconvolution. This approach is based on the Morozov discrepancy principle and on an exponential model function for the data term. The Total Variation image deconvolution is performed with the Alternating Direction Method of Multipliers (ADMM). With a smoothed l_2 norm, the differentiability of the value of the Lagrangian at the saddle point can be shown and an approximate model function obtained. The choice of the optimal parameter can be refined with a Newton method. The efficiency of the method is demonstrated on a blurred and noisy bone CT cross section.

1. Introduction

Linear inverse problems play a major role in image processing applications. For example, image deconvolution and reconstruction problems are encountered in medical imaging or image and video coding. Such problems can be described by the following linear relationship:

$$g^{\delta} = Af + w, \tag{1}$$

where g^{δ} represents the measured data for the noise level δ, f the ideal image, w an additive noise corrupting the measurements and A the blurring kernel. To solve this ill-posed problem and to obtain stable solutions in the presence of noise, a possible approach is to include some prior information and to minimize a regularization functional including a data fidelity term and a regularization term $R(f)$ that imposes an a priori constraint on the solution. A regularization parameter is balancing the contribution of the two terms. In the classical Tikhonov regularization, $R(f) = \|Df\|^2$, where D is a differential operator. More recently, the Total Variation regularization (TV) [1, 2] has been studied and applied to various image processing problems [3, 4]. For an image $f \in H_1(\Omega)$, this regularization is given by:

$$R_{TV}(f) = \int_{\Omega} |\nabla f(r)| dr \tag{2}$$

This regularization is known to preserve well edges. Various numerical methods have been used to solved the TV regularized deconvolution problem [5, 6]. Yet, extensive numerical experiments show that algorithms based on the Alternating Direction Method of Multipliers (ADMM) are the state-of-the-art methods [3, 4, 7, 8]. In order to obtain a good reconstructed image, it is necessary to choose an optimal regularization parameter. A very common approach is to try various values until the reconstruction solution is satisfactory. For large scale problems, an exhaustive search with a grid of points of regularization parameters values is very slow. Several methods have been studied for the optimal selection of the regularization parameter for Tikhonov regularization. Methods like the discrepancy principle and the Unbiased Predictive Risk Estimator are based on a knowledge of the noise level δ [2, 9]. The L-curve method and the Generalized Cross-Validation can be used without the knowledge of this noise level [10, 11]. Moreover, an iterative choice of the regularization parameters based on the Morozov discrepancy principle applied with a model function has been proposed in [12].

For TV regularization, fewer methods have been investigated. For denoising problems, Aujol et *al.* [13] have proposed a method where the denoised result is closed to the optimal, in the SNR sense. An extension of the Unbiased Predictive Risk Estimator has been shown to give a good estimate of the regularization parameter [14]. A method exploiting the Generalized Cross-Validation technique has been proposed in [15]. Babacan et *al* [16] have studied a bayesian variational method for solving TV deconvolution problems. A new approach based on a proximal point method applied to the dual formulation of the TV regularization problem and the discrepancy principle has been proposed recently [17].

The goal of this paper is to propose a fast and simple method to select the optimal regularization parameter for the restoration of an image f from a blurred and noisy image g^δ with TV regularization. Our contribution in this field is to design an iterative method following the idea of the model function of Kunish et *al.* [12]. The outline of this paper is the following. In Section 2, we present the linear inverse problem studied and the ADMM/TV algorithm chosen. In Section 3, the new method for the selection of the regularization parameter is presented. In Section 4, numerical results are given to demonstrate the effectiveness of the approach. Concluding remarks are given in Section 5.

2. Problem formulation, TV regularization and ADMM approach

2.1. Problem formulation

In this paper, we consider a linear problem of the form $g^\delta = Af + w$, where A is a linear and bounded convolution operator. We assume that the images f and g^δ are defined on a bounded open subset $\Omega \in \mathbb{R}^2$. The noisy data for the noise level δ is denoted as g^δ. This problem is ill-posed and to obtain a stable solution, we use a total variation regularization. The recent ADMM methodology is the state-of-the-art method to obtain a stable solution with the TV regularization [3, 4]. The convergence rates for the Augmented Lagrangian method when Morozov's discrepancy principle is chosen as stopping rule have been investigated in [18]. These type of algorithms have attracted much attention and have also been proposed to solve a number of image processing tasks, such as image impainting and deblurring. In the framework of the ADMM method, an augmented Lagrangian is considered given by:

$$\mathcal{L}(f, g_i, \lambda_i) = \sum(\|g_i\|_2 - \lambda_i^t(g_i - D_i f) + \tfrac{\beta}{2}\|g_i - D_i f\|_2^2) + \tfrac{\mu}{2}\|g^\delta - Af\|_{L_2}^2 \tag{3}$$

where μ is the regularization parameter, β the Lagrangian parameter, and $\|\|_2$ is the Euclidean norm in $\mathbb{R}^2$. For the ith pixel at the location (j, k), the first order finite difference operator D_i is defined by $D_i : \mathbb{R}^n \rightarrow \mathbb{R}^2$, $D_i(f) = (D_i^x(f), D_i^y(f))$, where D_i^x and D_i^y are the classical horizontal and vertical difference operators. In order to provide differentiability at the origin, the l_2 norm of $g_i = (g_i^1, g_i^2)$ is approximated by $\sqrt{(g_i^1)^2 + (g_i^2)^2 + \xi}$. The dual variable associated to the constraint $g_i = D_i(f)$ is denoted $\lambda_i \in \mathbb{R}^2$.

2.2. The ADMM algorithm and optimality conditions

The ADMM algorithm search for the saddle point of the augmented Lagrangian by iterating the following equations:

$$\begin{aligned} f^{k+1} &= \text{argmin}_x \mathcal{L}(x, g_i^k, \lambda_i^k) \\ g_i^{k+1} &= \text{argmin}_y \mathcal{L}(f^{k+1}, y, \lambda_i^k) \\ \lambda_i^{k+1} &= \lambda_i^k - \beta(g_i^{k+1} - D_i f^{k+1}) \end{aligned} \tag{4}$$

In this work, we have used the isotropic TV and the L_2 norm of the gradient. With the alternating minimization algorithm, three sequences (f^k, g^k, λ^k) are constructed with the following iterative scheme: For each pixel i

$$g_i^{k+1} = max\{\|D_i f^k + \frac{1}{\beta}(\lambda_i^k)\| - \frac{1}{\beta}, 0\} \frac{D_i f^k + \frac{1}{\beta}(\lambda_i^k)}{\|D_i f^k + \frac{1}{\beta}(\lambda_i^k)\|} \tag{5}$$

The new iterate f^{k+1} is obtained from the following linear system :

$$(\sum D_i^t D_i + \frac{\mu}{\beta} A^t A) f = \sum D_i^t (g_i^{k+1} - \frac{1}{\beta}\lambda_i^k) + \frac{\mu}{\beta} A^t g^\delta. \tag{6}$$

The Lagrange mutipliers are updated with, $\lambda_i^{k+1} = \lambda_i^k - \beta(g_i^{k+1} - D_i f^{k+1})$. The sequence $(f^k, g_i^k, \lambda_i^k)$ generated by the ADMM algorithm converges to a Kuhn-Tucker point $(f^*, g_i^*, \lambda_i^*)$ of the problem, which correspond to a saddle point of the Lagrangian. With the augmented Lagrangian given in (3), the Karush-Kuhn-Tucker (KKT) conditions can be written as:

$$\mu A^t (A f^*(\mu) - g^\delta) = -\sum D_i^t (\lambda_i^*), \quad \lambda_i^* \in \partial \|D_i f^*(\mu)\|_2, \quad g_i^* = D_i f^* \tag{7}$$

3. Iterative choice of the regularization parameter

In the following, we assume that the solution $f^*(\mu)$ obtained with the ADMM algorithm is unique. We first show the differentiability of $f^*(\mu)$ and of the value function $E(\mu)$ of the Lagrangian at the saddle point.

3.1. Lipschitz continuity and differentiability of $f^*(\mu)$

We first show that $f^*(\mu)$ is Lipschitz continuous. Subtracting the KKT condition for $\mu + t$ and μ yields:

$$\begin{aligned} \mu A^t A(f^*(\mu + t) - f^*(\mu)) \in t A^t A f^*(\mu + t) + t A^t g^\delta - \\ \sum D_i^t (\partial \|D_i f^*(\mu + t)\|_2 - \partial \|D_i f^*(\mu)\|_2). \end{aligned} \tag{8}$$

Taking the inner product with $f^*(\mu+t) - f^*(\mu)$, using the Schwarz inequality, that f^* is bounded and the monotonicity of the subdifferential, we obtain:

$$\mu \|A(f^*(\mu + t) - f^*(\mu))\|^2 \le C|t| \|f^*(\mu + t) - f^*(\mu)\| \tag{9}$$

where C is a constant. If A is injective, $\|A(f^*(\mu + t) - f^*(\mu))\|^2 \ge \alpha_{min} \|(f^*(\mu + t) - f^*(\mu))\|^2$, where α_{min} is the minimum singular value of $A^t A$ and thus f^* is Lipschitz continuous. We now show that the function $f^*(\mu)$ is differentiable and that its derivative is the unique solution ω of:

$$\mu A^t A \omega + \sum D_i^t \frac{D_i \omega}{\|D_i f^*(\mu)\|} = A^t g^\delta - A^t A f^*(\mu). \tag{10}$$

Let $\epsilon = (f^*(\mu + t) - f^*(\mu))/t - \omega$. Dividing (8) by t and substracting (10), we obtain:

$$\langle \mu A^t A \epsilon, \epsilon \rangle = -\langle A^t A (f^*(\mu + t) - f^*(\mu)), \epsilon \rangle + \sum \langle \gamma_i, \epsilon \rangle. \tag{11}$$

with $\langle \gamma_i, \epsilon \rangle = -\langle (\partial \|D_i f^*(\mu+t)\|_2 - \partial \|D_i f^*(\mu))\|_2)/t - \frac{D_i \omega}{\|D_i f^*(\mu)\|}, D_i \epsilon \rangle$.
By the Schwarz inequality, $\mu \alpha_{min} \|\epsilon\| \leq \|f^*(\mu+t) - f^*(\mu)\| + \sum \gamma_i$. When $t \to 0$, $\gamma_i \to 0$, and thus $\epsilon \to 0$ and $f^*(\mu)$ is thus differentiable.

3.2. Differentiability of the value function $E(\mu)$

In the following, we denote $E(\mu)$ the value obtained at the saddle point for $\mathcal{L}(f^*(\mu), g_i^*(\mu), \lambda^*(\mu), \mu)$. The differentiability of E follows from its definition and from the differentiability of $f^*(\mu)$. Classical differentiability calculation rules gives:

$$\partial E(\mu) = \mu \langle A f^*(\mu) - g^\delta, A f'^*(\mu) \rangle + \frac{\|g^\delta - A f^*(\mu)\|^2}{2} + \sum \langle D_i^t \partial \|D_i f^*(\mu)\|_2, f'^*(\mu) \rangle. \tag{12}$$

Taking into account the optimality condition, the first and the second derivative of $E(\mu)$ are given by:

$$E'(\mu) = \frac{\|g^\delta - A f^*(\mu)\|^2}{2}, \quad E''(\mu) = \langle A f^*(\mu) - g^\delta, A f'^*(\mu) \rangle \tag{13}$$

It can be shown that $E''(\mu) \geq 0$. The function $E(\mu)$ is monotonically increasing and convex.

3.3. The Morozov discrepancy principle and iterative choice of the regularization parameter

The Morozov discrepancy principle has been much studied in linear inverse problems [19]. In the Morozov discrepancy principle, the regularization parameter is chosen such that the discrepancy term is equal to the known noise level on the observed data δ, that is $\|g^\delta - A f^*(\mu)\| = c\delta$, for a constant $c \geq 1$. This condition can be written $2E'(\mu) = \delta^2$ for c=1.
In order to determine the regularization parameter and to solve the Morozov equation, we have used a two parameters model function. This model function and the approach detailed below for the choice of μ are independent of the method used to solve the Total Variation optimization problem. The approximate value obtained for the regularization parameter is refined with a Newton method. Following the model function methodology in [12], a model function can be calculated that preserve the main properties of $E'(\mu)$. Upon differentiating the optimality condition with respect to μ, we obtain:

$$A^t(A f^*(\mu) - g^\delta) + \mu A^t(A f'^*(\mu) - g^\delta) = -\frac{d}{d\mu} \sum D_i^t \frac{D_i f^*(\mu)}{\|D_i f^*(\mu)\|_2}. \tag{14}$$

Taking the inner product of the optimality condition with $f^*(\mu)$ and neglecting the second member, we obtain:

$$\langle (A f^*(\mu) - g^\delta), A f^*(\mu) \rangle + \mu \langle A f'^*(\mu) - g^\delta, A f^*(\mu) \rangle = 0 \tag{15}$$

This equation will be approximated with $E'(\mu) + \mu E''(\mu) = constant$. Solving this differential equation, we obtain a decreasing and convex two parameters model function $2E'(\mu) = a exp(-\mu) + b$ which is expected to describe the main properties of the value function $E'(\mu)$. This model function yields an approximate value of the regularization parameter. In a second step, we define the function $G(\mu)$ by $G(\mu) = 2E'(\mu) - \delta^2$. In order to solve $G(\mu) = 0$, given two initial guess values, μ_0, μ_1, a sequence of iterates $(\mu_k)_{k \geq 0}$ is generated with:

$$\mu_{k+1} = \mu_k - 0.1 \frac{2E'(\mu_k) - \delta^2}{E''(\mu_k)} \tag{16}$$

where the value $f'(\mu_k)$ is approximated with $\frac{f(\mu_k) - f(\mu_{k-1})}{\mu_k - \mu_{k-1}}$.

Journal of Physics: Conference Series **464** (2013) 012005 doi:10.1088/1742-6596/464/1/012005

4. Numerical experiments

In order to show the effectiveness of our method for choosing the regularization parameter, we have considered a simple deconvolution test problem. The image $f \in \mathbb{R}^{N^2}$ (N=328) to be recovered is a real synchrotron bone micro-CT slice displayed in Figure 1A. The image was artificially blurred with a Gaussian kernel of variance $\sigma_{blur}^2 = 1.46$ and a Gaussian noise with a variance σ^2=18.4 was added to the blurred image. The obtained image has the PSNR=24.87 and is displayed in Figure 1B. The noise level is thus set to $\delta^2 = N^2\sigma^2 = 1.98 \cdot 10^6$. In order

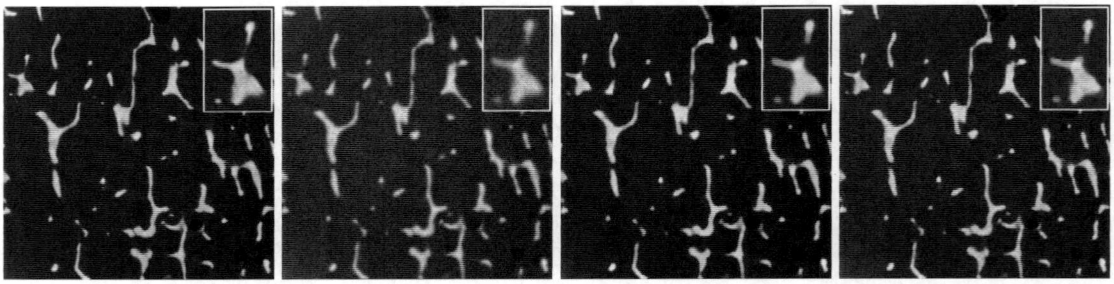

A. Ground truth image B. Blurred and noisy image C. Deconv. with $\mu_{Morozov}$ D. Deconv. with μ_{UPRE}

Figure 1.

to determine an approximate values of the parameters a and b of the model function $E'(\mu)$, two values $\mu_* = 0.2$ and $\mu_{**} = 10$ are first chosen. For each of the regularization parameters, μ_* and μ_{**}, the blurred noisy image is deconvolved with the ADMM algorithm and the values $E'(\mu_*)$ and $E'(\mu_{**})$ are calculated. The values obtained for a and b are $8.13 \cdot 10^5$ and $1.44 \cdot 10^6$ respectively. An approximate value for the regularization parameter, μ_{approx}=0.42, is obtained from the equation $\delta^2 = a exp(-\mu_{approx}) + b$. The Newton method is then applied to refine the solution. Starting from $\mu_0 = \mu_{approx}+0.1$ and $\mu_1 = \mu_{approx}$, the iterative formula (16) is applied. For each iteration of the Newton method, the ADMM algorithm is applied, 10 iterations being sufficient to obtain the convergence of the regularization functional. The Newton iterations are stopped when $|\frac{\|Af^*(\mu_k)-g^\delta\|^2-\delta^2}{\delta^2}| < 0.01$ The convergence is fast and only three iterations are necessary. The final value of the regularization parameter is $\mu_{Morozov} = 0.37$. The final image obtained with $\mu_{Morozov}$ has a PSNR=32.39 and is displayed in Figure 1C. The value obtained is a good estimate of the optimal value $\mu_{opt} = 0.8$ considering the PSNR of the reconstructed image, PSNR=33.03. The iterates μ_k have been reported on Figure 2-left, which displays the evolution of the data term as a function of the regularization parameter. An extensive comparison with other techniques to get the regularizing parameter will not be detailed here. Yet, the UPRE method for TV parameter selection [14] leads to the optimal regularization parameter μ_{UPRE}=1.4 (Figure 2-right) with a PSNR=32.45. The results obtained are thus rather similar in both cases but the proposed method does not require the calculation of the reconstructed images for all the regularization parameters.

5. Conclusion

In this paper, we proposed a new iterative method to determine the optimal regularization parameter in deconvolution problems with TV regularization based on the Morozov discrepancy principle. For each value of the regularization parameter, the TV regularization functional is minimized with the ADMM algorithm. An approximate model function for the data term of the regularization functional is first obtained. Then the solution is refined with a Newton iterative method. This method is very fast and avoid an extensive sweeping of the values of the regularization parameter.

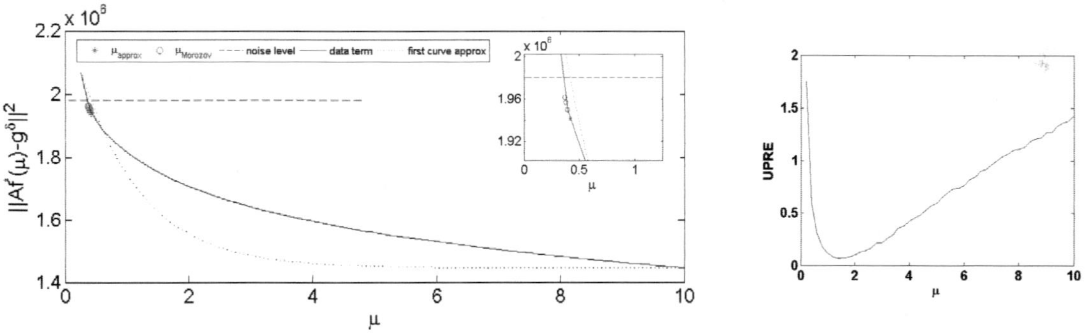

Figure 2. The Morozov principle (left); Unbiased Predictive Risk estimation (right)

References

[1] Rudin L I, Osher S and Fatemi E 1992 Nonlinear total variation based noise removal algorithms *Phys.* D **60** 259-68
[2] Vogel C R 2002 *Computational Methods for Inverse Problems* (Philadelphia, PA: SIAM)
[3] Wang Y, Yang J, Yin W and Zhang Y 2008 A new alternating minimization algorithm for total variation image reconstruction *SIAM Journal on Imaging Sciences* **1** 248-72
[4] Ng M K, Weiss P and Yuan X 2010 Solving constrained total-variation image restoration and reconstruction problems via alternating direction methods *SIAM Journal on Scientific Computing* **32** 2710-36
[5] Van den Berg E and Friedlander M P 2008 Probing the Pareto frontier for basis pursuit solutions *SIAM Journal on Scientific Computing* **31** 890-912
[6] Becker S, Bobin J and Candes E 2011 NESTA: A fast and accurate first-order method for sparse recovery *SIAM Journal on Imaging Sciences* **4** 1-39
[7] Yang J, Yin W, Zhang Y and Wang Y 2008 A fast algorithm for edge-preserving variational multichannel image restoration *SIAM Journal on Imaging Sciences* **2** 569-92
[8] Afonso M, Bioucas-Dias J and Figueiredo M 2011 An augmented Lagrangian approach to the constrained optimization formulation of imaging inverse problems *IEEE Transactions on Image Processing* **20** 681-95
[9] Morozov V A 1984 *Methods for Solving Incorrectly Posed Problems* ed Z Nashed (New York: Springer-Verlag)
[10] Golub G H, Heath M T and Wahba G 1979 Generalized cross-validation as a method for choosing a good ridge parameter *Technometrics* **21** 215-23
[11] Hansen P C and O'Leary D P 1993 The use of the L-curve in the regularization of discrete ill-posed problem *SIAM Journal on Scientific Computing* **14** 1487-503
[12] Kunish K and Zou J 1998 Iterative choices of regularization parameters in linear inverse problems *Inverse Problems* **14** 1247-64
[13] Aujol J F and Gilboa G 2006 Constrained and SNR-based solutions for TV-Hibert space image denoising *J. Math. Imaging Vis.* **26** 217-37
[14] Lin Y, Wohlberg B and Guo H 2010 UPRE method for total variation parameter selection *Signal Processing* **90** 2546-51
[15] Liao H, Li F and Ng M K 2009 Selection of regularization parameter in total variation restoration *JOSA* A **26** 2311-20
[16] Babacan S D, Molina R and Katsaggelos A K 2008 Parameter estimation in TV image restoration using variational distribution approximation *IEEE Transactions on Image Processing* **17** 326-39
[17] Wen Y W and Chan R H 2012 Parameter selection for total-variation-based image restoration using discrepancy principle *IEEE Transactions on Image Processing* **21** 1770-81
[18] Frick K, Lorenz D A and Resmerita E 2011 Morozov's principle for the augmented Lagrangian method applied to linear inverse problems *Multiscale Modeling and Simulation* 1528-48
[19] Engl H W, Hanke M and Neubauer A 2000 *Regularization of Inverse Problems* (Dordrecht: Kluwer Academic)

Journal of Physics: Conference Series **464** (2013) 012006 doi:10.1088/1742-6596/464/1/012006

Filter factor analysis of scaled gradient methods for linear least squares

Federica Porta, Anastasia Cornelio, Luca Zanni and Marco Prato

Department of Physics, Computer Science and Mathematics, University of Modena and Reggio Emilia

E-mail: `marco.prato@unimore.it`

Abstract. A typical way to compute a meaningful solution of a linear least squares problem involves the introduction of a filter factors array, whose aim is to avoid noise amplification due to the presence of small singular values. Beyond the classical direct regularization approaches, iterative gradient methods can be thought as filtering methods, due to their typical capability to recover the desired components of the true solution at the first iterations. For an iterative method, regularization is achieved by stopping the procedure before the noise introduces artifacts, making the iteration number playing the role of the regularization parameter. In this paper we want to investigate the filtering and regularizing effects of some first-order algorithms, showing in particular which benefits can be gained in recovering the filters of the true solution by means of a suitable scaling matrix.

1. Introduction

An ill-posed discrete inverse problem can be modeled with a linear system

$$\boldsymbol{b} = \boldsymbol{A}\boldsymbol{x}_{true} + \boldsymbol{\eta}, \tag{1}$$

where $\boldsymbol{A}$ is an ill-conditioned full-rank $m \times n$ matrix, with $m \geq n$, $\boldsymbol{b} \in \mathbb{R}^m$ is the observed data, $\boldsymbol{x}_{true} \in \mathbb{R}^n$ is the unknown object we want to recover and $\boldsymbol{\eta} \in \mathbb{R}^m$ is the noise corrupting the data. A classical example of the mathematical model (1) can be found in image restoration [2, 11]: in this case the matrix $\boldsymbol{A}$ represents the blurring effect that the acquisition process introduces on the unknown image $\boldsymbol{x}$.

Since discrete ill-posed problems have the property that the singular values decay to zero, a simple inversion of the data would amplify the noise providing a solution without any physical meaning. Especially when dealing with large scale problems, as the image deblurring one, a typical way to overcome this situation consists of iteratively building up a sequence of arrays that converges to the solution of the following least squares problem

$$\min_{\boldsymbol{x} \in \mathbb{R}^n} f(\boldsymbol{x}) \equiv \frac{1}{2}\|\boldsymbol{A}\boldsymbol{x} - \boldsymbol{b}\|_2^2, \tag{2}$$

and stopping the iterations before the noise distorts the reconstructions.

Due to the large diffusion of the least squares approach to address many real-world applications, the development of efficient methods to solve (2) has been widely treated in literature. In this paper, we consider the family of the scaled gradient methods, and we analyze the role played by

Journal of Physics: Conference Series **464** (2013) 012006 doi:10.1088/1742-6596/464/1/012006

well-known steplengths and scaling matrices in reconstructing the true solution. The starting point of our work is a paper of Nagy & Palmer [14], in which the authors investigate how the solution achieved by some classical gradient methods can be written as a linear combination of the singular vectors of $\boldsymbol{A}$ through appropriate filter factors [11]. Here we extend this analysis to the presence of a scaling matrix multiplying the gradient direction, showing its regularizing effect in recovering the true filter coefficients.
The paper is organized as follows: in Section 2 we introduce the optimization methods for the solution of (2) we decided to analyze, while in Section 3 the state-of-the-art on these algorithms as filtering methods is summarized, and the extension to the scaled case is provided. Section 4 will be devoted to some numerical experiments we carried out on a simulated image deblurring problem, while our conclusions are offered in Section 5.

2. Gradient methods

A gradient method for the solution of (2) is an iterative algorithm whose $(k+1)$–th element is defined by

$$\boldsymbol{x}_{k+1} = \boldsymbol{x}_k - \alpha_k \boldsymbol{M}_k \boldsymbol{g}_k, \tag{3}$$

where $\alpha_k > 0$ is the steplength, $\boldsymbol{M}_k$ is a symmetric and positive definite scaling matrix and $\boldsymbol{g}_k = \nabla f(\boldsymbol{x}_k) = \boldsymbol{A}^T(\boldsymbol{A}\boldsymbol{x}_k - \boldsymbol{b})$ is the gradient vector. The idea of finding a descent direction by scaling the gradient of the objective function is rather classical in numerical optimization. The most famous example of scaled gradient method is provided by the Newton algorithm, in which the scaling matrix is given by the inverse of the Hessian $\nabla^2 f(\boldsymbol{x}_k)$.
Among the large variety of selection rules for the steplength, classical examples are the Steepest Descent (SD) [5] and the Minimal Gradient (MG) [6, 15] parameters, which minimize $f(\boldsymbol{x}_k - \alpha \boldsymbol{M}_k \boldsymbol{g}_k)$ and $\|\nabla f(\boldsymbol{x}_k - \alpha \boldsymbol{M}_k \boldsymbol{g}_k)\|_2$, respectively:

$$\alpha_k^{SD} = \frac{\boldsymbol{g}_k^T \boldsymbol{M}_k \boldsymbol{g}_k}{\|\boldsymbol{A}\boldsymbol{M}_k \boldsymbol{g}_k\|_2^2} \quad ; \quad \alpha_k^{MG} = \frac{\boldsymbol{g}_k^T \boldsymbol{M}_k \boldsymbol{A}^T \boldsymbol{A} \boldsymbol{g}_k}{\|\boldsymbol{A}^T\boldsymbol{A}\boldsymbol{M}_k \boldsymbol{g}_k\|_2^2}. \tag{4}$$

In order to accelerate the slow convergence exhibited in most cases by the standard formulas (4), many other strategies for the steplength selection have been proposed, as the two Barzilai and Borwein rules [1] adapted to account for the scaling matrix [4], obtained by regarding the matrix $\boldsymbol{B}(\alpha_k) = (\alpha_k \boldsymbol{M}_k)^{-1}$ as an approximation of the Hessian $\nabla^2 f(\boldsymbol{x}_k)$ and forcing a quasi-Newton property on $\boldsymbol{B}(\alpha_k)$. The resulting steplengths are given by

$$\alpha_k^{BB1} = \frac{\boldsymbol{s}_{k-1}^T \boldsymbol{M}_k^{-1} \boldsymbol{M}_k^{-1} \boldsymbol{s}_{k-1}}{\boldsymbol{s}_{k-1}^T \boldsymbol{M}_k^{-1} \boldsymbol{y}_{k-1}} \quad ; \quad \alpha_k^{BB2} = \frac{\boldsymbol{s}_{k-1}^T \boldsymbol{M}_k \boldsymbol{y}_{k-1}}{\boldsymbol{y}_{k-1}^T \boldsymbol{M}_k \boldsymbol{M}_k \boldsymbol{y}_{k-1}},$$

where $\boldsymbol{s}_{k-1} = \boldsymbol{x}_k - \boldsymbol{x}_{k-1}$, $\boldsymbol{y}_{k-1} = \boldsymbol{g}_k - \boldsymbol{g}_{k-1}$, and their several alternated versions [9, 15].
As for the scaling matrices, in this paper we consider the following two examples arising from the constrained optimization:

- the one provided by the iterative space reconstruction algorithm (ISRA) [8], whose explicit expression of the scaling matrix is

$$\boldsymbol{M}_k^{ISRA} = \mathrm{diag}\left(\frac{\boldsymbol{x}_k}{\boldsymbol{A}^T\boldsymbol{A}\boldsymbol{x}_k}\right),$$

  where the quotient is intended in the Hadamard sense;

- the one proposed by Hager, Mair and Zhang (HMZ) [10], that exploits a gradient splitting strategy [12, 13] with a resulting scaling matrix given by

$$\boldsymbol{M}_k^{HMZ} = \text{diag}\left(\frac{\alpha_k^{CBB1}\boldsymbol{x}_k}{\boldsymbol{x}_k + \alpha_k^{CBB1}(\boldsymbol{A}^T\boldsymbol{A}\boldsymbol{x}_k + \boldsymbol{A}^T\boldsymbol{b})^+}\right),$$

where $t^+ = \max\{0, t\}$ and α_k^{CBB1} is a cyclic version of the first Barzilai and Borwein steplength rule computed by reusing α_k^{BB1} for p consecutive iterations [7].

For both matrices, the positive definiteness is ensured by thresholding the diagonal elements in a prefixed interval $0 < L_{\min} < L_{\max}$. We remark that here we are *not* considering the ISRA and HMZ algorithms, but we only borrow the scaling matrices defined in the algorithms themselves and use them in our scaled gradient scheme.

3. Filter factors analysis

Let $\boldsymbol{A} = \boldsymbol{U}\boldsymbol{\Sigma}\boldsymbol{V}^T$ be the singular value decomposition (SVD) of $\boldsymbol{A}$, where $\boldsymbol{U} = [\boldsymbol{u}_1\boldsymbol{u}_2\dots\boldsymbol{u}_m]$ and $\boldsymbol{V} = [\boldsymbol{v}_1\boldsymbol{v}_2\dots\boldsymbol{v}_n]$ are unitary matrices and $\boldsymbol{\Sigma} \in \mathbb{R}^{m\times n}$ is a diagonal matrix with entries $\sigma_1 \geq \sigma_2 \geq \dots \geq \sigma_n > 0$. When solving the linear ill-posed inverse problem (1), the generalized solution

$$\boldsymbol{x}^\dagger = \sum_{i=1}^{n} \frac{\boldsymbol{u}_i^T\boldsymbol{b}}{\sigma_i}\boldsymbol{v}_i = \boldsymbol{x}_{true} + \sum_{i=1}^{n} \frac{\boldsymbol{u}_i^T\boldsymbol{\eta}}{\sigma_i}\boldsymbol{v}_i$$

results to be totally useless due to the division by the small singular values and the consequent magnification of the corresponding noise components. A typical regularization strategy consists in simultaneously preserving the highest σ_i while filtering out the smaller ones by means of an array of positive weights φ_i. The corresponding regularized solution is given by

$$\boldsymbol{x}_{reg} = \sum_{i=1}^{n} \varphi_i \frac{\boldsymbol{u}_i^T\boldsymbol{b}}{\sigma_i}\boldsymbol{v}_i\,, \tag{5}$$

where φ_i are the so-called filter factors. The truncated singular value decomposition (TSVD) and the Tikhonov method are examples of this "direct" regularization approach, in which the filter factors are defined by

$$\varphi_i = \begin{cases} 1 & \text{if } i \leq r \\ 0 & \text{if } i > r \end{cases} \quad ; \quad \varphi_i = \frac{\sigma_i^2}{\sigma_i^2 + \lambda},$$

where $r \in \{1, \dots, n\}$ and $\lambda > 0$.

The general iteration of any gradient method can also be interpreted as a filtered regularized solution, since it can be written as

$$\boldsymbol{x}_{k+1} = \sum_{i=1}^{n} \varphi_i^{k+1} \frac{\boldsymbol{u}_i^T\boldsymbol{b}}{\sigma_i}\boldsymbol{v}_i\,,$$

where φ_i^{k+1} are opportune filter factors, depending on α_k and $\boldsymbol{M}_k$, automatically defined during the iterative procedure. In particular, if we assume $\boldsymbol{x}_0 = \boldsymbol{0}$, then:

- when $\boldsymbol{M}_k = \boldsymbol{I}$ for each iteration k, the iterative filter factors can be written as

$$\varphi_i^{k+1} = 1 - \prod_{\ell=0}^{k}\left(1 - \alpha_\ell\sigma_i^2\right), \qquad k = 0, 1, \dots \tag{6}$$

- when $\boldsymbol{M}_k$ is not trivial, the expression for φ_i^{k+1} is more complicated. In particular, once defined the polynomial P_k, acting on a $n \times n$ matrix $\boldsymbol{\Omega}$ as

$$P_k(\boldsymbol{\Omega}) = P_{k-1}(\boldsymbol{\Omega}) + \alpha_k \boldsymbol{M}_k(\boldsymbol{I} - \boldsymbol{\Omega} P_{k-1}(\boldsymbol{\Omega})), \quad P_{-1}(\boldsymbol{\Omega}) = \boldsymbol{0},$$

we have

$$\varphi_i^{k+1} = \sigma_i^2 \left(\frac{\sum_{j=1}^{n} (\boldsymbol{Q}_k)_{ij} \sigma_j \boldsymbol{u}_j^T \boldsymbol{b}}{\sigma_i \boldsymbol{u}_i^T \boldsymbol{b}} \right), \qquad k = 0, 1, \ldots \tag{7}$$

where $\boldsymbol{Q}_k = \boldsymbol{V}^T P_k(\boldsymbol{A}^T \boldsymbol{A}) \boldsymbol{V}$.

From equation (7) we can see that the presence of $\boldsymbol{M}_k$ on the filters expression makes any factor φ_i^{k+1} explicitly related to the whole singular system of $\boldsymbol{A}$, that acts in a different way on each i-th component of the filter vector. For the nonscaled case, this dependence is hidden in the α_ℓ coefficients, that are the same for all filters $\varphi_1^{k+1}, \ldots, \varphi_n^{k+1}$, while in the analytical expression only the i-th singular value σ_i is explicitly present. In Section 4, we will show the positive effect of such more complicated dependence in reconstructing the actual values of the true solution filter factors

$$\varphi_i^{true} = \sigma_i \frac{\boldsymbol{v}_i^T \boldsymbol{x}_{true}}{\boldsymbol{u}_i^T \boldsymbol{b}}. \tag{8}$$

4. Numerical experiments

The numerical experiments are carried out on a 2-dimensional image restoration test problem. We used the *phantom* test image from Matlab's Image Processing toolbox, artificially blurred by a Gaussian point spread function (PSF) with variance 1, in order to simulate the degrading effect on a real image due to the action of a general acquisition system. The resulting blurred data has been corrupted with 1% white Gaussian noise. We set $n = m = 4096$, thus obtaining $\boldsymbol{x}_{true}$ and $\boldsymbol{b}$ of 64×64 pixels. The first row of Figure 1 contains the true image, the PSF modeling the blurring effect and the measured data. When performing the scaled methods, we adopt the BB2 rule for the steplength selection, with a further Armijo reduction procedure to ensure the convergence.

Table 1 shows the best reconstruction errors $\|\boldsymbol{x}_k - \boldsymbol{x}_{true}\|_2 / \|\boldsymbol{x}_{true}\|_2$ for the methods we considered. For the HMZ matrix, we chose the value $p = 4$ for the cyclic version of the BB1 steplength rule defining $\boldsymbol{M}_k$. Moreover, the diagonal elements of the scaling matrices provided by the ISRA and HMZ approaches have been thresholded in the range $[10^{-3}, 10^8]$. From the results of Table 1, we can notice that the presence of the scaling matrix provides clear improvements in the minimum error reached by the corresponding methods, even if higher number of iterations are required. We have to remark that, for the scaled algorithms, the error behaviour rapidly decrease with the iterations and then stands in a very flat region, therefore values very close to the minimum can be obtained in both cases by stopping the procedure hundreds of iterations before the optimal value. In order to appreciate the positive effects of the scaling matrix on the reconstructions, the second row of Figure 1 reports the restored images by the different algorithms (since the reconstructions of the nonscaled methods were very similar, we reported only the BB2 case). In particular, we can observe that the presence of a scaling matrix helps to remove some artifacts due to the noise on the data and better recover some details of the solution. The better performances of the scaled algorithms are also visible in the plots of the filter coefficients; in particular, in Figure 2, we compare the filter factors generated at the iteration corresponding to the minimum error for the different nonscaled and scaled gradient methods and the "optimal" coefficients (8) associated to the true solution. The filter factors generated by the nonscaled method with different steplength exhibit smooth trends, which are not able to follow the (somewhat surprising) irregular profile of the filters related to $\boldsymbol{x}_{true}$, due to the presence of noise in the vector $\boldsymbol{b}$ at the denominator of the filters expression (8). On the contrary, the presence of the same denominator in the analytical form (7) allows a more faithful reconstruction of the filters by the scaled algorithms, whose behaviour results to be in general as erratic as that of the true ones.

Table 1. Numbers of iterations and minimum error reached for solving the image deblurring problem.

	Nonscaled gradient				Scaled gradient	
	MG	SD	BB1	BB2	ISRA+BB2	HMZ+BB2
Iter	1339	1331	195	204	1495	570
Min Err	0.2636	0.2636	0.2636	0.2636	0.1820	0.2258

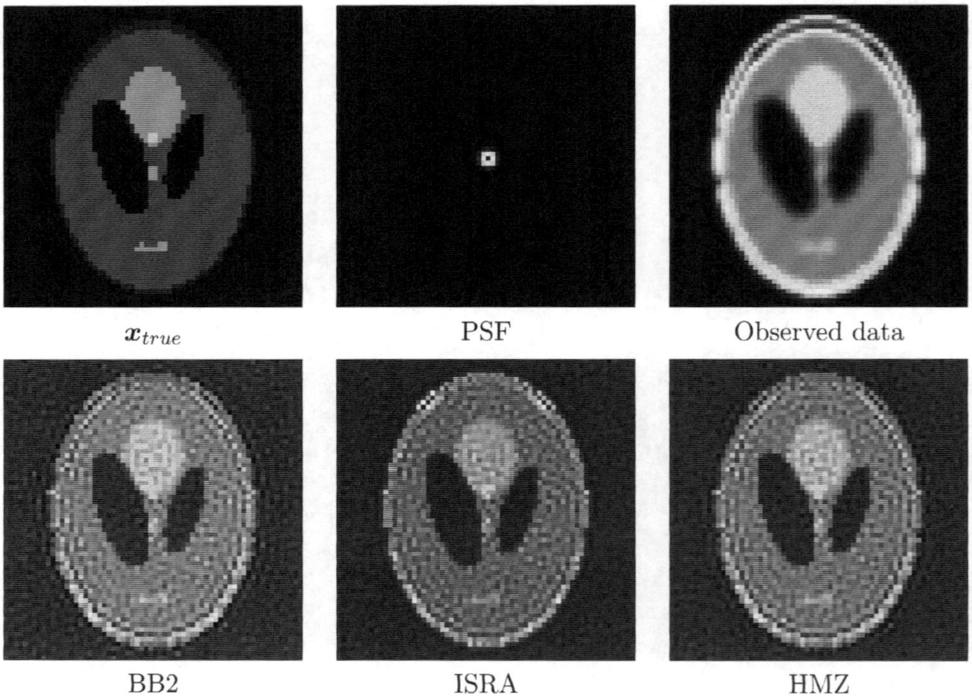

Figure 1. First row: true image, Gaussian PSF and measured data. Second row: reconstructions obtained with BB2 steplength and, from the left, $\boldsymbol{M}_k = \boldsymbol{I}$, $\boldsymbol{M}_k = \boldsymbol{M}_k^{ISRA}$, $\boldsymbol{M}_k = \boldsymbol{M}_k^{HMZ}$.

5. Conclusions

In this paper we conducted a comparative study between several iterative gradient methods for linear least squares problems, with the aim of investigating their ability in providing a regularized solution, i.e., a useful and stable solution not corrupted by the presence of noise on the data. In particular, the analysis we carried out regards the ability of a given scheme to reproduce faithfully the filter factors of the true solution. The filtering properties of nonscaled gradient methods had already been described by Nagy & Palmer: in our paper we generalized their considerations to the presence of a nontrivial scaling matrix in the descent direction. In particular, the formal

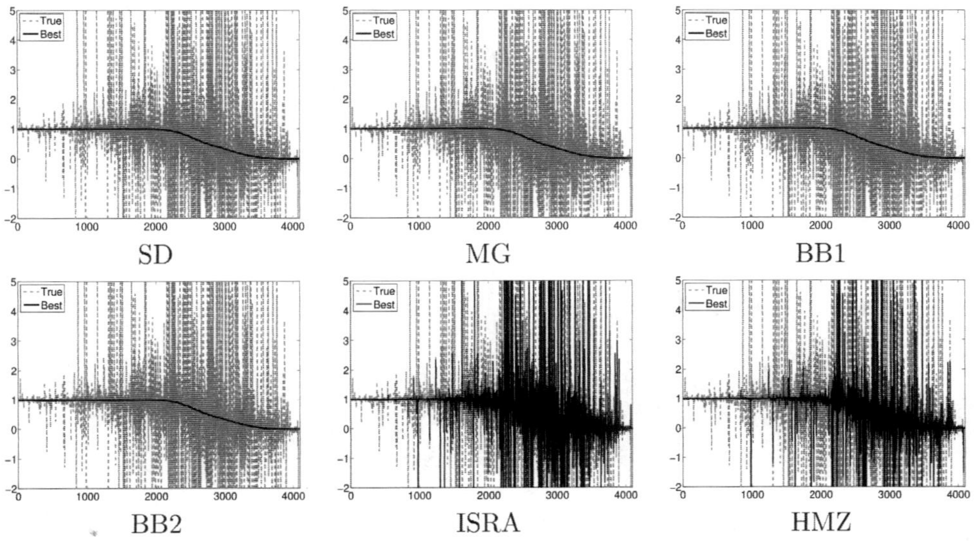

Figure 2. Comparison of the $\boldsymbol{x}_{true}$ filter factors (red dashed) with the ones generated by the gradient method with SD, MG, BB1, BB2 and the scaled (ISRA, HMZ) versions (black solid).

expression of the filter factors has been given and their numerical evaluation has been performed. We showed that the nonscaled methods are not able to generate filter factors that behave as erratically as those associated to the true solution. On the other hand, the scaled methods not only are able to recover the irregular profile of the exact filters, but also provide a better approximation of the unknown solution in terms of accuracy and goodness of the results (as remarked e.g. in [3]).

Acknowledgments

This work has been partially supported by the Italian Spinner2013 PhD Project "High-complexity inverse problems in biomedical applications and social systems" and by a grant of the Italian Gruppo Nazionale per il Calcolo Scientifico (GNCS) - Istituto Nazionale di Alta Matematica (INdAM).

References

[1] Barzilai J and Borwein J M 1988 *IMA J. Numer. Anal.* **8** 141–8
[2] Bertero M and Boccacci P 1998 *Introduction to inverse problems in imaging* (Bristol: Institute of Physics)
[3] Bonettini S Landi G Loli Piccolomini E and Zanni L 2013 *Int. J. Comput. Math.* **90** 9–29
[4] Bonettini S Zanella R and Zanni L 2009 *Inverse Probl.* **25** 015002
[5] Cauchy A 1847 *C. R. Acad. Sci. Paris* **25** 536–8
[6] Dai Y H and Yuan Y X 2003 *IMA J. Numer. Anal.* **23** 377-93
[7] Dai Y H Hager W W Schittkowski K and Zhang H 2006 *IMA J. Numer. Anal.* **26** 604–27
[8] Daube-Witherspoon M E and Muehllener G 1986 *IEEE T. Med. Imaging* **5** 61-6
[9] Frassoldati G Zanni L and Zanghirati G 2008 *J. Indust. Manag. Optim.* **4** 299–312
[10] Hager W W Mair B A and Zhang H 2009 *Math. Program.* **119** 1-32
[11] Hansen P C Nagy J G O'Leary D P 2006 *Deblurring images: matrices, spectra and filtering* (Philadelphia, PA: SIAM)
[12] Lantéri H Roche M Cuevas O and Aime C 2001 *Signal Process.* **81** 945–74
[13] Lantéri H Roche M and Aime C 2002 *Inverse Probl.* **18** 1397-419
[14] Nagy J and Palmer K 2003 *BIT* **43** 1003–17
[15] Zhou B Gao L and Dai Y H 2006 *Comput. Optim. Appl.* **35** 69-86

Journal of Physics: Conference Series **464** (2013) 012007 doi:10.1088/1742-6596/464/1/012007

Limit Cycles of Controlled Switched Systems: Existence, Stability, Sensitivity

L. Fribourg and R. Soulat
LSV - ENS Cachan & CNRS
E-mail: {fribourg,soulat}@lsv.ens-cachan.fr

Abstract. We present a control method which makes the trajectories of a switched system converge to a stable limit cycle lying in a desired region of equilibrium. The method is illustrated on the boost DC-DC converter example. We also point out in this example the sensitivity of limit cycles to parameter variations by showing how the limit cycle evolves in presence of small perturbations of some system parameters. This suggests that limit cycles are good candidates for reliable estimations of the physical parameters of switched systems using an appropriate inverse approach.

1. Introduction

Sampled switched systems are more and more used in electrical and mechanical industry, e.g. power electronics and automotive industry: this is due to their flexibility and simplicity for controlling accurately industrial mechanismes. These systems are governed by piecewise dynamics that are periodically sampled with a given period τ (see, e.g., [3]). At each sampling time, the "mode" of the system, i.e. the parameters of the dynamics, are switched according to a control rule. A classical example of sampled switched system is the boost DC-DC converter. The control of a switching cell puts the system, according to its position (open or closed), into one mode or another one. When the system stays in the same mode, it evolves towards a unique equilibrium point. In contrast, by adopting switching control rules, one can steer the system to a desired region R that is centered at a different equilibrium point, around which the system oscillates with some variability.

In [5], we have thus designed a state-dependent control method which decomposes R into a set of sub-boxes associated with specific control modes: at each sampling time, depending on the loacation of the system state, one activates a specific mode until the next sampling time. We show here that such a control not only ensures the invariance of the system state inside R, but makes the state trajectories converge to *limit cycles* inside R.

We also point out the sensitivity of limit cycles to parameter variations by showing the evolution of limit cycles in presence of small perturbations of system parameters. As indicated, e.g., in [9], this suggests that limit cycles are are good candidates for reliable estimation of physical parameters of the system

1.1. Related work

The presence of limit cycles in switched systems such as those used in power electronics has been often observed (see, e.g., [10]). The existence and stability of limit cycles is often shown using

Journal of Physics: Conference Series **464** (2013) 012007 doi:10.1088/1742-6596/464/1/012007

Lyapunov techniques (see, e.g., [2, 11]). An other common technique for proving existence of limit cycles is based on Poincaré map technique [7, 9]. Finally, sensibility functions w.r.t. initial conditions are used in order to compute a limit cycle [4], as well as describing functions [12].

We use here an original method showing that the existence of limit cycles follows from a property of piecewise invariance of a state space region, when the dynamics of the modes are contractive.

2. Decomposable Area of Interest

2.1. Model of sampled switched systems

Sampled switched systems, as defined, e.g., in [6], is a subclass of affine hybrid systems [8]. A sampled switched system Σ is defined by a triple $\langle \tau, U, \mathcal{F} \rangle$ where:

- $\tau \in \mathbb{R}_{\geq 0}$ is a "time sampling parameter",
- $U = \{1, \ldots, q\}$ is a finite set of modes,
- $\mathcal{F}$ is a set of the form $\{(A_u, B_u)\}_{u \in U}$ with $(A_u, B_u) \in \mathbb{R}^{n \times n} \times \mathbb{R}^{n \times 1}$.

We say that a piecewise $\mathcal{C}^1$ function $\mathbf{x} : \mathbb{R}_{\geq 0} \to \mathbb{R}^n$ is a *trajectory* of Σ if it is continuous and satisfies, for all $k \in \mathbb{N}$:

$$\exists u \in U \ \forall t \in [k\tau, (k+1)\tau) \quad \dot{\mathbf{x}}(t) = f_u \mathbf{x}(t),$$

with $(A_u, B_u) \in \mathcal{F}$. The trajectory is fully determined by the values $u_1, u_2, \ldots$ of u at sampling times $\tau, 2\tau, \ldots$. These values define a control function $\mathbf{u}(t)$, which is constant on each interval $[k\tau, (k+1)\tau)$, for all $k \in \mathbb{N}$. Between two sampling times, the system is governed by a differential equation of the form: $\dot{\mathbf{x}}(t) = A_u \mathbf{x}(t) + B_u$ with $u \in U$. We will use $\mathbf{x}(t, x, u)$ to denote the point reached by Σ at time t under mode u from the initial condition x. This defines a transition relation $\to_\tau^u$ given by: $x \to_\tau^u x'$ iff $\mathbf{x}(\tau, x, u) = x'$ for x and x' in $\mathbb{R}^n$.

NB: The expression $\mathbf{x}(\tau, x, u)$ can be put under the algebraic form: $(C_u \cdot x + D_u)$, with $C_u = e^{A_u \tau}$ and $D_u = e^{(A_u \tau - \mathcal{I}_n) A_u^{-1}} \cdot B_u$, where $\mathcal{I}_n$ denotes the identity matrix.

We define: $Post_u(X) = \{x' \mid x \to_\tau^u x' \text{ for some } x \in X\}$. We say that a zone X is *R-invariant via mode u* if $Post_u(X) \subseteq R$. We will group modes together into sequences called "patterns". Given a pattern π of the form $(u_1 \ldots u_m)$, we have:
$Post_\pi(X) = \{x' \mid x \to_\tau^{u_1} \cdots \to_\tau^{u_m} x' \text{ for some } x \in X\}$.
We say that X is *R-invariant via* π if $Post_\pi(X) \subseteq R$.

We say that R is "piecewise k-invariant" if it can be decomposed into a finite set of subsets that are R-invariant. Formally:

Definition 1 *A set $R \subseteq \mathbb{R}^n$ is* piecewise k-invariant *if there exists a finite set I of indices, a set $\{V_i\}_{i \in I}$ of subsets of $\mathbb{R}^n$ and a set $\{\pi_i\}_{i \in I}$ of patterns of length at most k such that:*

- $R = \bigcup_{i \in I} V_i$
- *V_i is R-invariant via π_i (i.e., $Post_{\pi_i}(V_i) \subseteq R$), for all $i \in I$.*

In practice, R and V_i ($i \in I$) are *boxes*, i.e., cartesian products of closed intervals of $\mathbb{R}^n$. The set $\{(V_i, \pi_i)\}_{i \in I}$ is said to be a *k-invariant decomposition* of R.

Given a decomposition Δ of the from $\{V_i, \pi_i)\}_{i \in I}$, we define $Post_\Delta$ as follows:

$$Post_\Delta(X) = \bigcup_{i \in I} Post_{\pi_i}(X \cap V_i), \quad \text{for all } X \subset R.$$

In [5], we have proposed a method in order to show that, given a set R, it is piecewise k-invariant by constructing a k-invariant decomposition of R. This method is recalled in Appendix.

Journal of Physics: Conference Series **464** (2013) 012007 doi:10.1088/1742-6596/464/1/012007

In the rest of this paper, we will suppose that we are given a piecewise k-invariant set R and a k-invariant decomposition $\Delta = \{(V_i, \pi_i)\}_{i \in I}$. The decomposition Δ induces a state-dependent control that makes any trajectory starting fom R go back to R within at most k steps: Given a starting state x_0 in R, we know that $x_0 \in V_i$ for some $i \in I$ (since $R = \bigcup_{i \in I} V_i$); one thus applies π_i to x_0, which gives a new state x_1 that belongs to R (since V_i is R-invariant via π_i); the process is repeated on x_1, and so on iteratively. Formally, for any $x_0 \in R$, the above process leads to the construction of a sequence of points $\{x_i\}_{i \geq 0}$ of R, and a sequence of indices $\{k_i\}_{i \geq 0}$ defined, for all $i \geq 0$, by:

- $x_{i+1} = Post_{\pi_{k_i}}(x_i)$
- $k_i = argmin\{j \in I \mid x_i \in V_j\}$.

In the following, we will suppose that every mode of U is "contractive". We say that a mode $u \in U$ is *contractive* if there exists ρ_u such that $0 < \rho_u < 1$ and, for all $x, y \in \mathbb{R}^n$: $\|Post_u(x) - Post_u(y)\| \leq \rho_u \|x - y\|$.

2.2. Boost DC-DC Converter

Let us illustrate the model of sampled switched system and the result of the decomposition procedure (see Appendix).

Example 1 *This example is taken from [1] (see also, e.g., [6, 13]). This is a boost DC-DC converter with one switching cell (see Fig. 1). There are two operation modes depending on the position of the switching cell. An example of pattern of length 4 is illustrated on Fig. 2: it corresponds to the application of mode 2 on $(0, \tau]$ and mode 1 on $(\tau, 4\tau]$. The controlled system is steered to a zone where the output voltage stabilizes around a desired value. The state of the system is $x(t) = [i_l(t), v_c(t)]^T$ where i_l is the current intensity in inductor, and $v_c(t)$ the voltage of capacitor C. The dynamics associated with mode u is of the form $\dot{x}(t) = A_u x(t) + b_u$ ($u = 1, 2$) with*

$$A_1 = \begin{pmatrix} -\frac{r_l}{x_l} & 0 \\ 0 & -\frac{1}{x_c}\frac{1}{r_0 + r_c} \end{pmatrix} \quad b_1 = \begin{pmatrix} \frac{v_s}{x_l} \\ 0 \end{pmatrix}$$

$$A_2 = \begin{pmatrix} -\frac{1}{x_l}(r_l + \frac{r_0 . r_c}{r_0 + r_c}) & -\frac{1}{x_l}\frac{r_0}{r_0 + r_c} \\ \frac{1}{x_c}\frac{r_0}{r_0 + r_c} & -\frac{1}{x_c}\frac{1}{r_0 + r_c} \end{pmatrix} \quad b_2 = \begin{pmatrix} \frac{v_s}{x_l} \\ 0 \end{pmatrix}$$

We will use the numerical values of [1], expressed in the per unit system: $x_c = 70$, $x_l = 3$, $r_c = 0.005$, $r_l = 0.05$, $r_0 = 1$, $v_s = 1$. The sampling period is $\tau = 0.5$. We can check that modes 1 and 2 are contractive (i.e., $\|e^{A_u \tau}\| = \rho_u < 1$ for $u = 1, 2$). For the equilibrium region, we take $R = [1.55, 2.15] \times [1.0, 1.4]$, which corresponds to a medium value 1.85 for i_l with ± 0.3 for variability, and medium value 1.20 for v_c with ± 0.2 for variability.

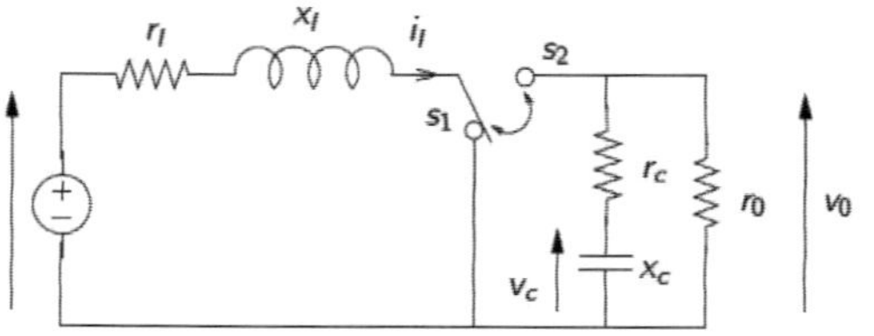

Figure 1. Scheme of the boost DC-DC converter

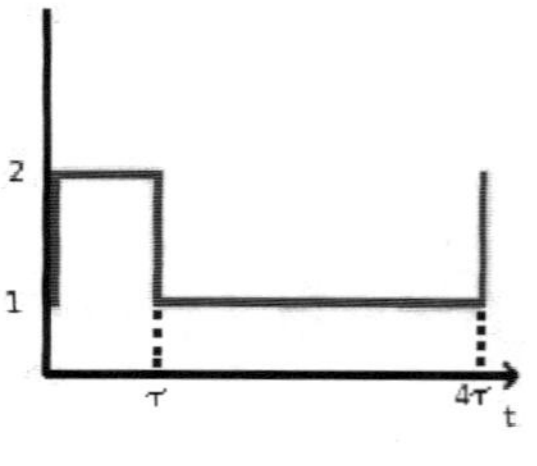

Figure 2. Switching pattern (2.1.1.1)

The application of the algorithm Decomposition (see Appendix) to $R = [1.55, 2.15] \times [1.0, 1.4]$, *with* $k = 5$ *and* $d = 1$, *yields a decomposition of* $R = \bigcup_{j=1}^{4} V_j$ *depicted in Figure 3. The associated patterns are:* $\pi_1 = (11222)$, $\pi_2 = (2)$, $\pi_3 = (212)$, $\pi_4 = (1)$, *with:* $V_1 = [1.55, 1.85] \times [1.00, 1.20]$, $V_2 = [1.85, 2.15] \times [1.00, 1.20]$, $V_3 = [1.85, 2.15] \times [1.20, 1.40]$, $V_4 = [1.55, 1.85] \times [1.20, 1.40]$.

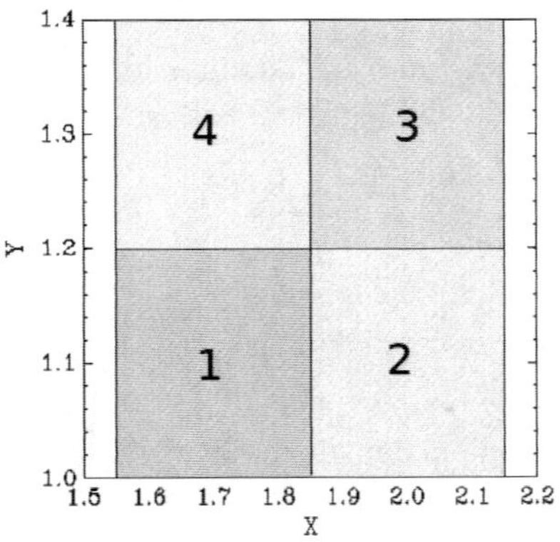

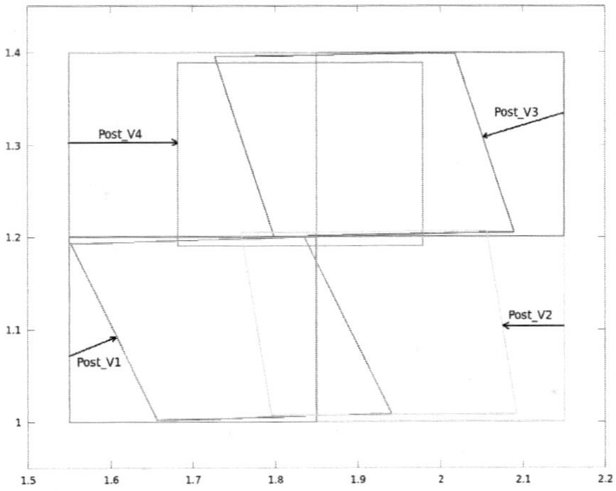

Figure 3. Left: Decomposition of R into 4 subsets. Right: visualization of R-invariance ($Post_{\pi_i}(V_i) \subset R$)

In Figure 4, we depict runs that start from the four corners of R, and follows the control strategy induced by this decomposition . In all cases, one observes a phenomenon of convergence of the trajectories to a unique limit cycle. This phenomenon follows from the existence of the decomposition of R, as explained in next Section.

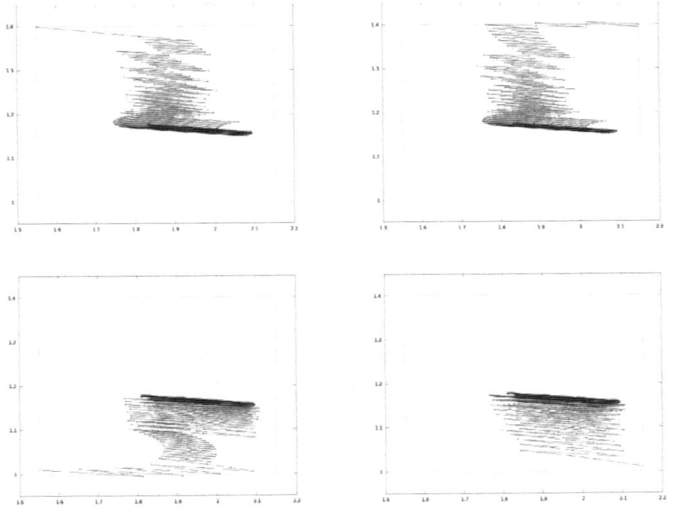

Figure 4. Runs starting from the four corners of R, following the control strategy induced by the decomposition

3. Limit Cycles

3.1. Existence and stability

If all the modes are contractive (i.e., $\|e^{A_u\tau}\| = \rho_u < 1$ for all $u \in U$), the existence of a k-invariant decomposition $\Delta : \{(V_i, \pi_i)\}_{i \in I}$ of R entails the existence of stable limit cycles to which any trajectory $\mathcal{T}$ issued from R converges provided that all the points of $\mathcal{T}$ remain at a minimum distance $e > 0$ (of arbitrary positive value) from any border of a sub-box V_i $(i \in I)$. Formally, we have:

Theorem 1 *Consider a trajectory $\mathcal{T} : \{x_0, x_1, x_2, \cdots\}$ of points of R induced by Δ, which all stay at a minimum distance $e > 0$ from the borders of V_i. Then there exists a set of points $\mathcal{C} = \{y_0, y_1, \cdots, y_{m-1}\}$ of R, and an integer N such that, for all $j \in \{0, \cdots, m-1\}$:*

- $Post_\Delta(y_j) = y_{j+1}$ *(where y_m stands for y_0),*
- $Post_\Delta(x_{N+j+i.m}) = x_{N+j+i.m+1}$,
- $\lim_{i\to\infty} x_{N+j+i.m} = y_j$.

The proof of the above statement (see Appendix 5) shows also that $\mathcal{C}$ is a stable limit cycle for the trajectories of R, in the sense that a small perturbation occurring in the trajectory $\mathcal{T}$ does not prevent the subsequent trajectory to converge to $\mathcal{C}$. For the boost converter, the corresponding limit cycle is depicted in Figure 5.

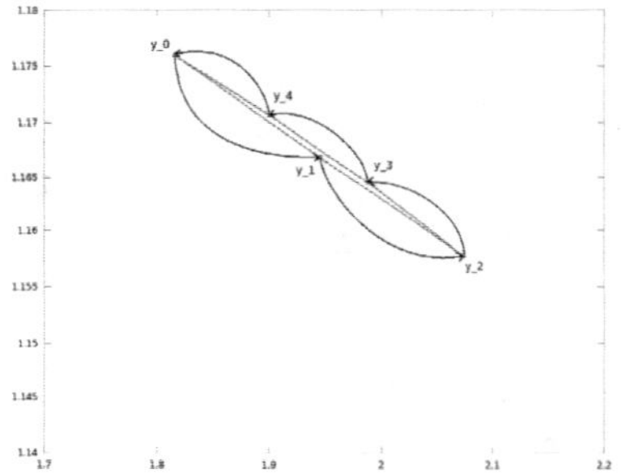

Figure 5. Limit cycle for $r_0 = 1$ (corresponding to pattern π_1 : (11222)): one has $y_0 \to^1 y_1 \to^1 y_2 \to^2 y_3 \to^2 y_4 \to^2 y_0$

3.2. Sensitivity

Given the region R and fixed values for the physical parameters of the system, the decomposition procedure gives a decomposition $\Delta : \{(V_i, \pi_i)\}_{i \in I}$. Actually, even in presence of small variations of parameters of the system, the same decomposition of Δ can ensure the piecewsie invariance of R. For example, the decomposition Δ found for the boost DC-DC converter for $r_0 = 1$ and $R = [1.55, 2.15] \times [1.0, 1.4]$ (see Figure 3), still ensures the piecewise invariance of R when r_0 varies from 0.975 to 1.005. This guaranteed that the control found $r_0 = 1$ makes the system converge to a stable limit cycle of R, for small variations of r_0. However, as shown in Figure 6, the form and position of the limit cycle is very sensitive to the actual value of r_0:

For $r_0 = 0.965$, the limit cycle corresponds to repeated pattern $(\pi_3\pi_1^3\pi_3\pi_1^2\pi_3\pi_1^3\pi_3\pi_1^3\pi_3\pi_1^3)$ (with $\pi_1 = (11222)$, $\pi_3 = (212)$); for $r_0 = 0.975$, the associated pattern is $(\pi_3\pi_1^5)$, while, for $r_0 = 1$ and $r_0 = 1.005$, the pattern is (π_1).

This suggests that limit cycles are good candidates for reliable estimations of the physical parameters of switched systems, using an appropriate inverse approach.

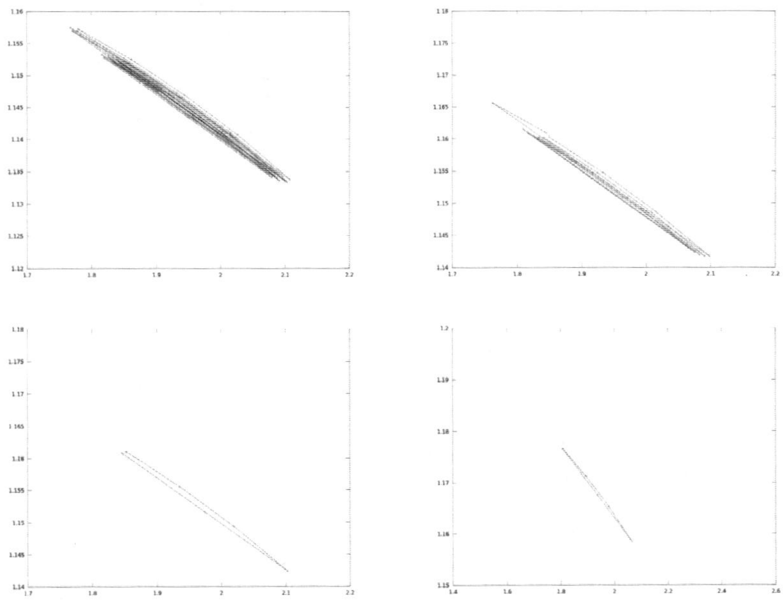

Figure 6. Limit cycles for $r_0 = 0.965$ (pattern $(\pi_3\pi_1^3\pi_3\pi_1^2\pi_3\pi_1^3\pi_3\pi_1^3\pi_3\pi_1^3)$) on the upper left, for $r_0 = 0.975$ (pattern $(\pi_3\pi_1^5)$) on the upper right, for $r_0 = 1$ (pattern π_1) on the lower left, and for $r_0 = 1.005$ (pattern π_1) on the lower right

4. Final Remarks

We have shown that, for controlled switched systems, the existence of an invariance decomposition for a given zone of interest, entails the convergence of the system to a stable limit cycle within the zone.

We have also shown the high sensitivity of the form and position of limit cycles when some physical parameter of the system slighty varies, such as load resistance in DC-DC boost converters. This suggests that limit cycles are good candidates for reliable estimations of the physical parameters of switched systems,

5. References

[1] A. Beccuti, G. Papafotiou, and M. Morari. Optimal control of the boost dc-dc converter. In *Proc. 44th IEEE Conference on Decision and Control European Control Conference (CDC-ECC '05)*, pages 4457 – 4462, dec. 2005.

[2] J. Buisson, P.-Y. Richard, and H. Cormerais. On the stabilisation of switching electrical power converters. In *HSCC*, volume 3414 of *LNCS*, pages 184–197. Springer, 2005.

[3] J. Cámara, A. Girard, and G. Gößler. Safety controller synthesis for switched systems using multi-scale symbolic models. In *Proc. 50th IEEE Conference on Decision and Control (CDC-ECC '11)*, pages 520–525, 2011.

[4] D. Flieller, P. Riedinger, and J.-P. Louis. Computation and stability of limit cycles in hybrid systems. *Nonlinear Analysis*, 64(2):352–367, 2006.

[5] L. Fribourg and R. Soulat. Controlled recurrent subspaces for sampled switched linear systems. Research Report LSV-12-24, LSV, ENS Cachan, France, Dec 2012.

[6] A. Girard, G. Pola, and P. Tabuada. Approximately bisimilar symbolic models for incrementally stable switched systems. *IEEE Trans. on Automatic Control*, 55:116–126, 2010.

[7] J. M. Gonçalves. Region of stability for limit cycles of piecewise linear systems. In *IEEE Conference on Decision and Control, Maui*, 2003.

[8] T. A. Henzinger. The theory of hybrid automata. In *Proceedings of the 11th Annual IEEE Symposium*

on Logic in Computer Science, LICS '96, pages 278–292, Washington, DC, USA, 1996. IEEE Computer Society.

[9] I. Hiskens. Inverse problems in power systems. In *Bulk Power System Dynamics and Control V*, pages 180–195, 2001.

[10] D. Patino Guevara. *Pilotage des cycles limites dans les system es dynamiques hybrides. Application aux alimentations electriques statiques.* Ph.D Thesis, Nancy-université, 2009.

[11] M. Rubensson, , M. Rubensson, and B. Lennartson. Stability of limit cycles in hybrid systems using discrete-time lyapunov techniques. In *Proceedings of the 39th IEEE Conference on Decision and Control*, 2000.

[12] S. Sanders. On limit cycles and the describing function method in periodically switched circuits. *IEEE Trans. Circuits and Systems*, 40(9):564–572, 1993.

[13] M. Senesky, G. Eirea, and T.-J. Koo. Hybrid modelling and control of power electronics. In *HSCC*, volume 2623 of *Lecture Notes in Computer Science*, pages 450–465, 2003.

Appendix: Decomposition Procedure [5]

Suppose we are given a box $R \subseteq \mathbb{R}^n$. We now give a procedure in order to show that R is piecewise k-invariant. If we find a pattern π of length at most k such that R is invariant, then we are done. Otherwise, we decompose R dichotomously into 2^n sub-boxes $V_1, \cdots, V_{2^n}$ of equal size. If for each box V_i, we find a pattern π_i of length at most k such that V_i is R-invariant via π_i, we are done. If, for some V_j, no such pattern exists, we recursively apply the process to V_j. The process ends with success when a k-invariant decomposition of R is found, or with failure when a maximal degree d of decomposition has been reached. The algorithmic form of this method is given in Algorithms 1 and 2. For the sake of simplicity, we have considered the case of space dimension $n = 2$ (the extension to the general case is straightforward). The procedure Decomposition(W,R,D,K) is called with R as input value for W, d for input value for D, and k as input value for K. The procedure Decomposition(W,R,D,K) calls the subprocedure Find_Pattern(W,R,K) in order to find a pattern for which W is R-invariant. This is simply done by testing all the patterns by increasing length order, until one is found that maps W into R. The correctness of the decomposition procedure is stated formally as follows ([5]): If the procedure Decomposition(R,R,d,k) terminates with success (i.e: $\Delta \neq False$), then it returns a k-invariant decomposition Δ of R.

Algorithm 1: Decomposition(W,R,D,K)

Input: A box W, a box R, a degree D of decomposition, a length K of pattern
Output: Δ which is either a K-invariant decomposition of W or *False*

$\pi := Find\_Pattern(W, R, K)$
if $\pi \neq False$ **then**
  return $\Delta := (W, \pi)$
else
  if $D = 0$ **then**
    return $\Delta := False$
  else
    Dichotomously decompose W into (W_1, W_2, W_3, W_4) /* (case $n = 2$) */
    $\Delta_1 :=$ Decomposition(W_1,R,$D-1$,K)
    $\Delta_2 :=$ Decomposition(W_2,R,$D-1$,K)
    $\Delta_3 :=$ Decomposition(W_3,R,$D-1$,K)
    $\Delta_4 :=$ Decomposition(W_4,R,$D-1$,K)
    if $\Delta_1 = False \vee \Delta_2 = False \vee \Delta_3 = False \vee \Delta_4 = False$ **then**
      return $\Delta := False$
    else
      return $\Delta := \Delta_1 \cup \Delta_2 \cup \Delta_3 \cup \Delta_4$

Appendix: Sketch of Proof of Theorem 1

Let us assume that we have a decomposition $\Delta = \{(V_i; \pi_i)\}_{i \in I}$ of R. Let us consider a trajectory $\mathcal{T} = \{x_0, x_1, \cdots\}$, where all x_i stay at minimum distance $e > 0$ from the borders of Δ. For every point x_i of $\mathcal{T}$, we consider the ball B_i centered at x_i of radius e. Clearly, every B_i is included in $\mathring{V}_j$ for some $j \in I$. Furthermore we have $Post_\Delta(B_i) \subset Bi + 1$, since all the modes of U are contractive. We claim that there exists $\varepsilon > 0$ such that: either $\lambda(B_{i+1} \cap \overline{\bigcup_{j \leq i} B_j})$ or $B_{i+1} \subset \bigcup_{j \leq i} B_j$. The first subcase cannot occur indefinitely, since eventually all R is covered in a finite number of steps. Hence, the second subcase occurs eventually ($B_{i+1} \subset \bigcup_{j \leq i} B_j$), this entails the existence of a cycle passing by a point of B_{i+1}.

Algorithm 2: Find_Pattern(W,R,K)

```
Input: A box W, a box R, a length K of pattern
Output: π wich is either a pattern mapping W into R or False
for i = 1...K do
    Π := set of patterns of length i
    while Π is non empty do
        Select π in Π
        Π := Π \ {π}
        if Post_π(W) ⊆ R then
            return π
return π := False
```

An image reconstruction method from Fourier data with uncertainties on the spatial frequencies

Anastasia Cornelio[1], **Silvia Bonettini**[2] **and Marco Prato**[1]

[1] Dipartimento di Scienze Fisiche, Informatiche e Matematiche, Università di Modena e Reggio Emilia, Via Campi 213/b, 41125 Modena, Italy

[2] Dipartimento di Matematica e Informatica, Università di Ferrara, Via Saragat 1, 44122 Ferrara, Italy

E-mail: marco.prato@unimore.it

Abstract. In this paper the reconstruction of a two-dimensional image from a nonuniform sampling of its Fourier transform is considered, in the presence of uncertainties on the frequencies corresponding to the measured data. The problem therefore becomes a blind deconvolution, in which the unknowns are both the image to be reconstructed and the exact frequencies. The availability of information on the image and the frequencies allows to reformulate the problem as a constrained minimization of the least squares functional. A regularized solution of this optimization problem is achieved by early stopping an alternating minimization scheme. In particular, a gradient projection method is employed at each step to compute an inexact solution of the minimization subproblems. The resulting algorithm is applied on some numerical examples arising in a real-world astronomical application.

1. Introduction

Fourier imaging finds applications in several scientific and medical areas, as computerized tomography, magnetic resonance imaging and astronomy [4, 9]. More in general, a Fourier approach is adopted when the high frequencies characterizing the measured radiation make the acquisition systems designed for optical wavelengths unsuitable. In this paper, we address the Fourier image reconstruction problem in which the frequencies corresponding to the data samples are not exactly known, but some estimates of them are available. Problems in which the degraded image has to be found without prior knowledge of the system PSF are typically known as blind deconvolution [17, 18, 19]. In particular, in our case we are dealing with a semi-blind deconvolution problem [2, 10, 11], since the PSF is not completely unknown but a parametrized version is available. Due to the usual non-negativity of the image and the partial a priori information on the frequencies, the reconstruction problem can be reformulated as a constrained nonlinear least squares problem. We propose to address its solution by means of a cyclic block gradient projection scheme, in which at each cycle either the image or the parametrized PSF is kept fixed, while the other is updated through a certain number of gradient projection iterations. Moreover, a suitable criterion allows us to perform only few cycles, thus leading to a regularized solution thanks to the semiconvergent behaviour of a general iterative approach. The proposed scheme is then validated with some numerical tests on simulated datasets, designed within a real-world astronomical application.

Journal of Physics: Conference Series **464** (2013) 012008 doi:10.1088/1742-6596/464/1/012008

2. Imaging from Fourier data

The Fourier imaging problem aims at reconstructing a n^2-vector f, representing the discretization of an unknown non-negative distribution $f(x, y)$ (rearranged in lexicographic order), starting from the knowledge of an N-vector g containing the corresponding measured or estimated complex Fourier samples. Mathematically, this can be done by solving the minimization problem

$$\min_{f \in \mathbb{R}^{n^2},\ f \geq 0} \quad \Phi(f) \equiv \frac{1}{2} \|Af - g\|^2_{\mathbb{C}^N}, \tag{1}$$

where the $N \times n^2$ matrix A is defined as

$$(A)_{hk} = e^{2\pi i \rho_k (x_h \cos(\theta_k) + y_h \sin(\theta_k))}, \quad h = 1, \ldots, n^2, \quad k = 1, \ldots, N. \tag{2}$$

Here (θ_k, ρ_k) are the spatial frequencies corresponding to the sample g_k, given in polar coordinates (in several applications, like the one described in Section 3, the data acquisition system makes a polar representation of the spatial frequency more convenient than the classical Cartesian one).

If the data frequencies are not completely known, but only some ranges of variability are available, then problem (1) becomes

$$\min_{\substack{f \in \mathbb{R}^{n^2},\ f \geq 0 \\ \theta \in \mathbb{R}^N,\ \theta^{\min} \leq \theta \leq \theta^{\max} \\ \rho \in \mathbb{R}^N,\ \rho^{\min} \leq \rho \leq \rho^{\max}}} \quad \Phi(f, \theta, \rho) \equiv \frac{1}{2} \|A(\theta, \rho) f - g\|^2_{\mathbb{C}^N}, \tag{3}$$

where $\theta = (\theta_1, ..., \theta_N)^T$ and $\rho = (\rho_1, ..., \rho_N)^T$ are the vectors of angular and radial coordinates, and $\theta^{\min}$, $\theta^{\max}$, $\rho^{\min}$, $\rho^{\max}$ are the related lower and upper bounds.

Problem (3) is convex if restricted to f only, but is nonconvex with respect to θ and ρ and, even more, with respect to (f, θ, ρ), thus leading to the possible presence of several local minima.

Due to the separate nature of the optimization variables, we approached the minimization problem (3) by means of an alternating strategy [13, 14], consisting in the following iterative minimization scheme:

$$f^{(\ell+1)} = \underset{f \geq 0}{\operatorname{argmin}}\ \Phi(f, \theta^{(\ell)}, \rho^{(\ell)}) \tag{4}$$

$$\theta^{(\ell+1)} = \underset{\theta^{\min} \leq \theta \leq \theta^{\max}}{\operatorname{argmin}}\ \Phi(f^{(\ell+1)}, \theta, \rho^{(\ell)}) \tag{5}$$

$$\rho^{(\ell+1)} = \underset{\rho^{\min} \leq \rho \leq \rho^{\max}}{\operatorname{argmin}}\ \Phi(f^{(\ell+1)}, \theta^{(\ell+1)}, \rho) \tag{6}$$

The key problem is that the proposed scheme has two nontrivial drawbacks, since a) the convergence to a solution of (3) is not guaranteed (see e.g. the counterexample shown by Powell in [25]), and b) computing the exact minimum points in each subproblem is impractical. Both problems can be avoided by solving inexactly each partial minimization problem by means of a suitable descent method [5, 13]. In particular, this occurs when the approximation of each solution of (4)–(6) is achieved by performing a finite number of iterations of the gradient projection (GP) method [7], that applies to any minimization problem over a convex set Ω of the form

$$\min_{x \in \Omega} \psi(x)$$

and whose main steps are reported in Algorithm 1.

The GP method and its scaled version SGP [7] combine a gradient projection step with variable steplength (and scaling) with the well-known Armijo rule to achieve the sufficient decrease in the

Journal of Physics: Conference Series **464** (2013) 012008 doi:10.1088/1742-6596/464/1/012008

Algorithm 1 Gradient projection method

Choose the starting point $x^{(0)} \in \Omega$, set the parameters $\nu, \mu \in (0,1)$, $0 < \alpha_{min} < \alpha_{max}$.

FOR $k = 0, 1, 2, \ldots$ DO THE FOLLOWING STEPS:

STEP 1. Choose the steplength parameter $\alpha_k \in [\alpha_{min}, \alpha_{max}]$;
STEP 2. Projection: $y^{(k)} = P_\Omega(x^{(k)} - \alpha_k \nabla\psi(x^{(k)}))$;
STEP 3. Descent direction: $\Delta x^{(k)} = y^{(k)} - x^{(k)}$;
STEP 4. Backtracking loop: set $\lambda_k = 1$.
IF $\psi(x^{(k)} + \lambda_k \Delta x^{(k)}) \leq \psi(x^{(k)} + \nu\lambda_k \nabla\psi(x^{(k)})^T \Delta x^{(k)}$ THEN go to Step 5;
ELSE set $\lambda_k = \mu\lambda_k$ and go to Step 4.
ENDIF
STEP 5. Set $x^{(k+1)} = x^{(k)} + \lambda_k \Delta x^{(k)}$.

END

Algorithm 2 Cyclic block gradient projection method

Choose $f^{(0)} \geq 0$, $\theta^{\min} \leq \theta^{(0)} \leq \theta^{\max}$, $\rho^{\min} \leq \rho^{(0)} \leq \rho^{\max}$ and three integers $N_f, N_\theta, N_\rho \geq 1$.

FOR $\ell = 0, 1, 2, \ldots$ DO THE FOLLOWING STEPS:

STEP f. Choose an integer $1 \leq N_f^{(\ell)} \leq N_f$ and compute $f^{(\ell+1)}$ by applying $N_f^{(\ell)}$ iterations of Algorithm 1 to problem (4) starting from the point $f^{(\ell)}$.
STEP θ. Choose an integer $1 \leq N_\theta^{(\ell)} \leq N_\theta$ and compute $\theta^{(\ell+1)}$ by applying $N_\theta^{(\ell)}$ iterations of Algorithm 1 to problem (5) starting from the point $\theta^{(\ell)}$.
STEP ρ. Choose an integer $1 \leq N_\rho^{(\ell)} \leq N_\rho$ and compute $\rho^{(\ell+1)}$ by applying $N_\rho^{(\ell)}$ iterations of Algorithm 1 to problem (6) starting from the point $\rho^{(\ell)}$.

END

objective function [3]. As concerns the choice of the steplength parameter, notable results have been obtained in denoising and deblurring problems [21, 26] by alternating the two Barzilai-Borwein rules [1, 12]. More details on GP and its parameters can be found e.g. in [6, 24].
In conclusion, problem (3) can be approximately solved by applying to the minimum problems (4)–(6) a finite number of GP steps: the resulting scheme is the cyclic block gradient projection (CBGP) method detailed in Algorithm 2 and any limit point of the sequence $(f^{(\ell)}, \theta^{(\ell)}, \rho^{(\ell)})$ is stationary (see [5, Theorem 4.2]).

3. An application in astronomy

In this section we test the proposed scheme in a real-world application in astronomy, namely the reconstruction of X-ray images of a solar flare. The Fourier samples of the unknown image are estimated from a set of count profiles provided by the solar satellite RHESSI [20], launched by NASA on February 5 2002 to study solar flares and other energetic solar phenomena. Due to the RHESSI imaging hardware, based on rotating modulation collimator techniques [15], the data frequencies are arranged around nine concentric circles in the Fourier plane whose radii form a geometric sequence with a common ratio $\sqrt{3}$ (see Figure 1). In particular, the total number of data N is given by the sum of the N_i samples on each circle ($i = 1, \ldots, 9$).
The resulting mathematical model is given by

$$\min_{\substack{f \in \mathbb{R}^{n^2},\ f \geq 0 \\ \theta \in \mathbb{R}^N,\ \theta^{\min} \leq \theta \leq \theta^{\max}}} \quad \Phi(f, \theta) \equiv \frac{1}{2} \|A(\theta) f - g\|^2_{\mathbb{C}^N}, \tag{7}$$

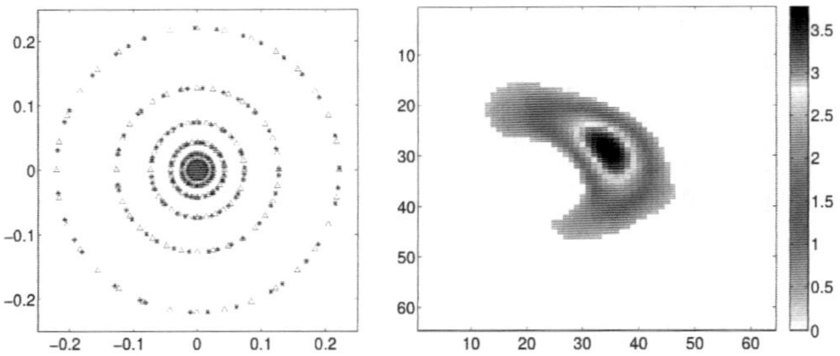

Figure 1. *Left panel*: example of distribution of the sampling points for the RHESSI imaging problem (black asterisks), together with the bounds of the box constraints $\theta^{\min}, \theta^{\max}$ (red triangles). *Right panel*: image of a real flare used as ground truth for the simulated datasets.

that is a particular case of (3) in which the radial coordinate ρ of each sample is known. The CBGP approach is compared with the Space-D algorithm [6], that is a GP method developed within the RHESSI data analysis framework for solving the minimum problem (1). As all the other Fourier-based reconstruction algorithms designed for RHESSI [8, 22], Space-D assumes that, for the i-th circle ($i = 1, \dots, 9$), the angular coordinate θ is given by the middle points of a uniform partition of the $[0, 2\pi]$ range in N_i bins, since in most cases this choice provides a good estimate of the real value. Our CBGP approach allows to remove this assumption, and to treat the value of θ as a further unknown of the reconstruction problem, using the uniform discretization to define the box constraints (see Figure 1).

We generated a simulated dataset by selecting an image of a real flare (April 15 2002, 00:05:00–00:10:00 UT, energy band 12-14 keV - see Figure 1, right panel) and calculating the corresponding data through numerical integration of the Fourier Transform. As concerns the choice of θ, we chose a random angle within each box $[\theta^{\min}, \theta^{\max}]$. The Fourier samples in this dataset will be characterized by the presence of a systematic source of noise only, due to the fact that in the inversion procedure the exact value of θ is unknown. We refer to this dataset with the term "Sim". Finally, we corrupted the resulting data by realistic statistical noise (estimated by assuming Poisson noise on the original count profiles), and this last dataset will be denoted by "Sim_N". We point out that our way to generate the simulated dataset is a very simplified version of the much more sophisticated procedure employed by RHESSI. We chose this procedure since it allows to know the "true" values of θ and, therefore, to evaluate the effectiveness of the blind approach. In particular, the choice of angles θ according to a uniform random distribution is arbitrary.

As for the CBGP parameters, we used the same values selected for the Space-D algorithm: in particular, the stopping criterion chosen to terminate the iterations in both steps of the cyclic approach[1] is given by

$$|\psi(x^{(k)}) - \psi(x^{(k-1)})| < 10^{-4}|\psi(x^{(k)})|. \tag{8}$$

The choice of the threshold 10^{-4} guarantees a limited number of iterations at each step and, therefore, a regularization effect on the recovered solution (more details are shown in [7]). The CBGP scheme is initialized with a vector $\theta^{(0)}$ equal to the middle points of each bin. In these settings, the image reconstructed by the Space-D algorithm is exactly $f^{(1)}$ (see Algorithm 2). As far as the number of cycles ℓ concerns, the CBGP algorithm is stopped when criterion (8) in either the step on f or the step on θ is satisfied at the first iteration (in addition, a maximum number of cycles equal to 20 is imposed).

[1] We remark that in this situation both the block–iterative optimization methods described in Section 2 reduce to the alternate inexact solution of subproblems (4)–(5).

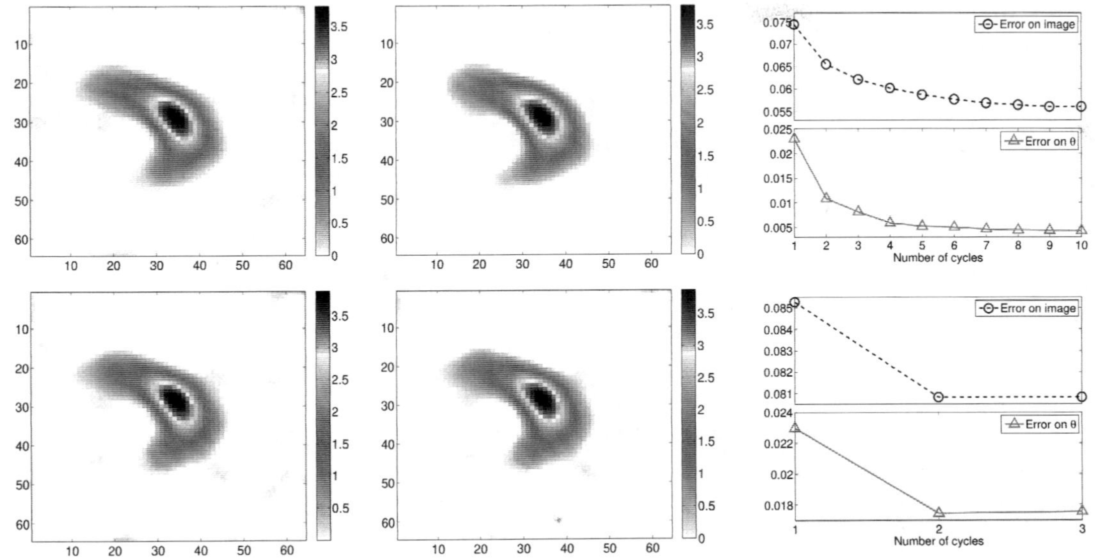

Figure 2. Results of the simulated tests Sim (top row) and Sim_N (bottom row). The images obtained with Space-D (first column) and CBGP (second column) are presented, together with the reconstruction errors on the image and, for CBGP, on the angular coordinates (third column).

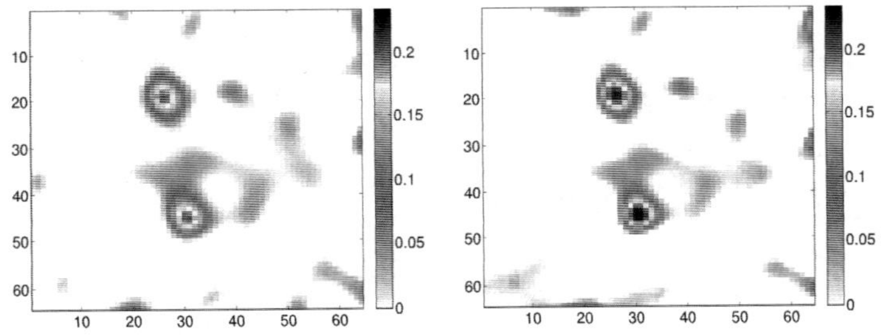

Figure 3. Reconstructions obtained from real data with Space-D (left panel) and CBGP (right panel).

In Figure 2 we report the reconstructed images obtained with Space-D (first column) and CBGP (second column), together with the relative errors in Euclidean norm on both the image and the angular coordinates (third column), for the Sim (top row) and Sim_N datasets (bottom row). As remarked before, the reconstruction error provided by Space-D is the one obtained at the first cycle. From the results obtained we can see that the blind approach is able to improve the image quality with respect to the Space-D algorithm, thanks to a better estimate of the underlying array θ. Although the restored images appear to be very similar, some details are better reconstructed, as the right footpoint of the loop in the Sim case. The lower number of cycles performed in the noisy case is in agreement with the fact that, in presence of statistical noise, the effect of a bad choice for θ is more negligible (even if still present). We remark that, for the considered application, the computational time is not a crucial issue, since both Space-D and CBGP end in few seconds.
Fore sake of completeness, we also considered a real event occurred on February 20, 2002 (11:06:02–11:06:34 UT, energy band 22-26 keV) and we show the reconstructions obtained with Space-D and CBGP in Figure 3.

4. Conclusions and future work

In this work we considered the image reconstruction problem from a sampling of its Fourier Transform, in the specific case in which only estimates of the frequencies corresponding to the measured data are available. From the mathematical point of view, the presence of uncertainties on the frequencies can be translated in a constrained least-squares problem, in which the matrix that describes the link between data and image is not known a priori, but depends on a vector of parameters to be determined. The natural subdivision of the unknowns in separate groups led us to choose a cyclic block alternating method for the resolution of the optimization problem. In particular, we used a strategy in which each subproblem is solved inexactly by means of a projected gradient method, which on the one hand provides a convergent scheme, and on the other hand allows us to avoid the (onerous) exact resolution of each subproblem. As numerical experiments, we chose to test the method on a simulated dataset inspired by a real application in astronomy, and the results seem to validate the proposed scheme.

Future work will involve the extension of the semi-blind approach to objective functions including regularization penalties on the image, like the Tikhonov term for smooth objects or an edge-preserving function for sharp contents. As regards the astronomical application considered in this work, we would like to extend the proposed methodology to the reconstruction of electrons images through the Bremsstrahlung integral equation [16, 23].

Acknowledgements

This work has been partially supported by the FIRB - Futuro in Ricerca research project “Apprendere nel tempo: un nuovo approccio computazionale per l’apprendimento automatico di sistemi dinamici”, contract RBFR12M3AC_002, and the research project “Metodi numerici e software per l’ottimizzazione su larga scala con applicazioni all’image processing” of the Italian Gruppo Nazionale per il Calcolo Scientifico (GNCS-INdAM).

References

[1] Barzilai J and Borwein J M 1988 *IMA J. Numer. Anal.* **8** 141–8
[2] Benvenuto F and Ferrari A 2010 *Inverse Probl.* **26** 105011
[3] Bertsekas D P 1999 *Nonlinear programming* (Belmont, MA: Athena Scientific)
[4] Blahut R E 2001 *Theory of Remote Image Formation* (Cambridge: Cambridge University Press)
[5] Bonettini S 2011 *IMA J. Numer. Anal.* **37** 1431–52
[6] Bonettini S and Prato M 2010 *Inverse Probl.* **26** 095001
[7] Bonettini S Zanella R and Zanni L 2009 *Inverse Probl.* **25** 015002
[8] Bong S C Lee J Gary D E and Yun H S 2006 *Astrophys. J.* **636** 1159–65
[9] Bracewell R 2000 *The Fourier Transform and its applications* (New York, NY: McGraw-Hill)
[10] Conan J M Mugnier L M Fusco T Michau V and Rousset G 1998 *Appl. Opt.* **37** 4614–22
[11] Christou J C Bonnacini D Ageorges N and Marchis F 1999 *Messenger* **97** 14–22
[12] Frassoldati G Zanni L and Zanghirati G 2008 *J. Indust. Manag. Optim.* **4** 299–312
[13] Grippo L and Sciandrone M 1999 *Optim. Method Softw.* **10** 587–637
[14] Grippo L and Sciandrone M 2000 *Oper. Res. Lett.* **26** 127–36
[15] Hurford G J et al 2002 *Solar Phys.* **210** 61–86
[16] Kontar E P et al 2011 *Space Sci. Rev.* **159** 301–55
[17] Kundur D and Hatzinakos D 1996 *IEEE Signal Proc. Mag.* **13**(3) 43–64
[18] Kundur D and Hatzinakos D 1996 *IEEE Signal Proc. Mag.* **13**(6) 61–63
[19] Levin A Weiss Y Durand F and Freeman W T 2009 *IEEE Conf. Computer Vision and Pattern Recognition* pp 1964–71
[20] Lin R P et al 2002 *Solar Phys.* **210** 3–32
[21] Loris I Bertero M De Mol C Zanella R and Zanni L 2009 *Appl. Computat. Harmon. A.* **27** 247–54
[22] Massone A M et al 2009 *Astrophys. J.* **703** 2004–16
[23] Prato M 2009 *Arch. Comput. Method E.* **16** 109–60
[24] Prato M Cavicchioli R Zanni L Boccacci P and Bertero M 2012 *Astron. Astrophys.* **539** A133
[25] Powell M J D 1973 *Math. Program.* **4** 193–201
[26] Zanella R Boccacci P Zanni L and Bertero M 2009 *Inverse Probl.* **25** 045010

Journal of Physics: Conference Series **464** (2013) 012009 doi:10.1088/1742-6596/464/1/012009

Depth-variant blind restoration with pupil-phase constraints for 3D confocal microscopy

Saima Ben Hadj [1], Laure Blanc-Féraud [1], and Gilbert Engler [2]

[1] Morphème research group - I3S/INRIA/CNRS/IBV/UNS,
2000, route des Lucioles, 06903 Sophia Antipolis Cedex, France
[2] Unit IBSV, INRA, 06903 Sophia Antipolis, France

E-mail: Saima.Ben_Hadj@inria.fr, Laure.Blanc_Feraud@inria.fr, Gilbert.Engler@sophia.inra.fr

Abstract. Three-dimensional images of confocal laser scanning microscopy suffer from a depth-variant blur, due to refractive index mismatch between the different mediums composing the system as well as the specimen, leading to optical aberrations. Our goal is to develop an image restoration method for 3D confocal microscopy taking into account the blur variation with depth. The difficulty is that optical aberrations depend on the refractive index of the biological specimen. The depth-variant blur function or the Point Spread Function (PSF) is thus different for each observation. A blind or semi-blind restoration method needs to be developed for this system. For that purpose, we use a previously developed algorithm for the joint estimation of the specimen function (original image) and the 3D PSF, the continuously depth-variant PSF is approximated by a convex combination of a set of space-invariant PSFs taken at different depths. We propose to add to that algorithm a pupil-phase constraint for the PSF estimation, given by the the optical instrument geometry. We thus define a blind estimation algorithm by minimizing a regularized criterion in which we integrate the Gerchberg-Saxton algorithm allowing to include these physical constraints. We show the efficiency of this method relying on some numerical tests.

1. Introduction

Confocal laser scanning microscopy allows to observe in three dimensions (3D) biological living specimens, at a resolution of few hundred nanometers. However, 3D images coming from this system are mainly affected by two types of distortion: a depth-variant (DV) due to the refractive index variation between the different mediums composing the system as well as the specimen, and a Poisson noise due to the photon counting process at the sensor. The knowledge of the Point Spread Function (PSF) is crucial for the restoration of these images. Theoretical models for the 3D PSF computation were previously developed in the literature based on the geometrical optics. For example, we use the model described in [3, 5] for the simulation of blurred confocal microscopy images. Nevertheless, this model relies on the knowledge of some physical acquisition parameters which are inaccessible in practice such as the refractive index of the biological specimen. The estimation of the 3D DV PSF is an important problem which is highly under-determined.

In our previous work, we considered an approximation of the 3D DV PSF by a convex combination of a set of space-invariant (SI) PSFs taken at different depths [1]. A method for the joint estimation of the specimen function (the original image) and the 3D DV PSF is proposed in [1] and summarized in section 2.

In this article, we propose to add an additional constraint on the PSF coming from modeling the optical instruments. In fact, the PSF of confocal microscopy can be computed from the magnitude of a complex function known as the *complex valued-amplitude PSF* [3, 5]. The shape and the support of this complex function are given in the Fourier domain by the numerical aperture of the optical system (this latter is known since it is fixed by the system manufacturer) [5]. We propose to incorporate these constraints in the joint image and PSF estimation algorithm presented in [1]. For that, we use the Gerchberg-Saxton algorithm [4] which allows to impose these constraints on the PSF (i.e. constraints on the pupil phase).

This article is organized as follows : We first recall the DV PSF model and the joint estimation algorithm proposed in [1]. We then describe the constraints to be imposed on the estimated PSF, arising from the optic modeling. We present the proposed algorithm for incorporating these constraints in the estimation algorithm. We finally present some numerical results showing the interest of these constraints.

2. Joint image and PSF estimation algorithm

Let $f,\, g : \mathcal{I} \subset \mathbb{N}^3 \to \mathbb{R}^+$ be two functions denoting respectively the original and the degraded discrete 3D images ($\mathcal{I}$ being the image support). The forward degradation model is given by the following Poisson statistics: $g \sim \mathcal{P}\left(\widetilde{H}\,(f) + b_g\right)$ where $b_g > 0$ is a strictly positive constant modeling the background noise, coming from the specimen auto-fluorescence or light scattering. b_g is assumed to be known since it can be estimated from a dark region which does not contain the object. The DV blur operator $\widetilde{H}$ can be modeled as follows [1]: $\widetilde{H}\,(f) = \sum_{1 \le i \le M} h_i * (\psi_i.f)$ where h_i, $i = 1, ..., M$ is a set of SI PSFs taken at different depths and $\psi_i : \mathcal{I} \to [0, 1]$, $i = 1, ..., M$ is a set of weighting functions such that $\sum_{1 \le i \le M} \psi_i(x, y, z) = 1, \forall (x, y, z) \in \mathcal{I}$. The choice of these weighting functions is discussed in [3] where it is shown that piecewise-linear functions are sufficient to obtain a good restoration. We thus consider that $\psi_i\,(x, y, z)$ are varying along (OZ) axis and constant along (OX) and (OY) axis. For fixed weighting functions, the image and the PSFs can be estimated by minimizing the following criterion [1]:

$$
\begin{aligned}
J\left(f,\, h^1,\, ...,\, h^M\right) = & \sum_{(x,y,z)\in\mathcal{I}} \left[\widetilde{H}\,(f)\,(x, y, z) - g(x, y, z)\, log\, \left(\widetilde{H}\,(f)\,(x, y, z) + b_g\right)\right] \\
& + \alpha \left\|\nabla f\right\|_1 + \sum_{1 \le i \le M} \beta_i \left\|\nabla h_i\right\|_2^2
\end{aligned}
\tag{1}
$$

The first term is the data fidelity term arising from the Poisson statistic. The second term defines the total variation for the image smoothing while preserving edges. The last term promotes regular and spread PSFs in order to capture all the blur of the observed image in the PSF. α and β_i, $i = 1, ..., M$ are regularization parameters. We define the set of admissible images as follows: $C_f = \{f; f \ge 0;\, \|f\|_1 = \|g - b_g\|_1\}$ and the set of admissible PSFs as follows: $C_h = \{h; h \ge 0; \|h\|_1 = 1; supp\,(h) \subset B\}$, $B \subset \mathcal{I}$ being the PSF support. The joint image and PSF estimation can be performed as in [1] by alternatively minimizing the criterion (1) w.r.t f and $h_i, i = 1, ..., M$, the constraints of C_f and C_h being involved using a fast scaled gradient projection (SGP) [2]:

(i) Object estimation using SGP [2]: $\hat{f}^{(k+1)} = \arg\min_{f \in C_f} J\left(f,\, \hat{h1}^{(k)},\, ...,\, \hat{hM}^{(k)}\right)$

(ii) PSF estimation using SGP [2]: For $i = 1, ..., M$, $\hat{h_i}^{(k+1)} = \arg\min_{h_i \in C_h} J\left(\hat{f}^{(k+1)},\, \hat{h1}^{(k)},\, h_i,\, ...,\, \hat{hM}^{(k)}\right)$

It is worth stressing that the minimized criterion is non-convex w.r.t all the variables $(f, h^1, ..., h^M)$ but it is convex w.r.t. each of the variables separately. Thus the proposed algorithm seeks for a local minimum which depends on the PSF initialization. In our tests, the PSFs are initialized from the theoretical PSF model that we present in the following section by setting approximately its physical parameters (namely the refractive index of the immersion medium and the sample).

3. Pupil phase constraints for the PSF estimation

Since the problem of blind restoration is under-determined, it is important to inject as much information as possible on the unknown variables. In particular, the PSF comes from modeling the optical instrument and can be written as follows [3, 5]:

$$h_i(x, y, z) = |h_{Ai}(x, y, z)|^4 \tag{2}$$

where h_{Ai} is a complex function $h_{Ai}(x, y, z) \in \mathbb{C}$ called *complex valued-amplitude PSF* or *coherent PSF*. In that model, excitation and emission wavelengths are assumed to be sufficiently close in such a way that their associated coherent PSFs could be considered of the same magnitude. Moreover, the pinhole diameter is assumed to be sufficiently small in such a way that it could be modeled by a Dirac distribution and thus does not appear in the PSF model (2). Furthermore, we have :

$$h_{Ai}(x, y, z) = TF_{2D}^{-1}(P_i(k_x, k_y, z)) \tag{3}$$

where $P_i(k_x, k_y, z) \in \mathbb{C}$ is the 2D pupil function computed for each Z-slice, TF_{2D}^{-1} is the 2D inverse Fourier transform, and k_x, k_y are the 2D frequency coordinates. This pupil function can be decomposed into two terms:

$$P_i(k_x, k_y, z) = P_{ai}(k_x, k_y)\, P_d(k_x, k_y, z) \tag{4}$$

The term $P_d(k_x, k_y, z)$ models the defocus of the optical instruments and is assumed to be known since it is given by the system physical parameters [5]. The term $P_{ai}(k_x, k_y) \in \mathbb{C}$ models the aberrations affecting the wavefront when traveling through the different system mediums. It is unknown since it is related to the biological specimen. Note that the aberration term does not depend on the z coordinate while the defocus term depends on that coordinate, which allows us to introduce constraints on the PSF in the Fourier domain. The support C of the aberration term is a disk of radius given by the numerical aperture (NA) of the microscope: $C = \{(k_x, k_y) \in \mathbb{N}^2;\ \sqrt{k_x^2 + k_y^2} < \frac{2\pi}{\lambda} NA\}$. This model allows us to impose additional constraints on the PSF (shape and support).

4. Blind restoration with pupil phase constraints

To take into account the constraints given in section 3, we add to the previous algorithm (section 2) an additional step allowing to estimate the complex function h_{Ai}, by estimating in the Fourier domain the aberration term P_{ai}, the defocus term P_d is fixed by the acquisition system. In order to estimate the term P_{ai} from the PSF h_i, we propose to use the Gerchberg-Saxton algorithm which was successfully applied in [4] to wide field microscopy. The global estimation algorithm is the following, k being the iteration counter:

(i) Object estimation using SGP algorithm [2]: $\hat{f}^{(k+1)} = \arg\min_{f \in C_f} J\left(f, \hat{h^1}^{(k)}, ..., \hat{h^M}^{(k)}\right)$

(ii) PSF estimation: For $i = 1, ..., M$, iterate the following steps:

(a) Estimation of the PSF using SGP [2]: $\hat{h_i}^{(k+1)} = \arg\min_{h_i \in C_h} J\left(\hat{f}^{(k+1)}, \hat{h1}^{(k)}, h_i, ..., \hat{hM}^{(k)}\right)$

(b) Estimation of the aberration term ${P_{ai}}^{(k+1)}$ from the PSF $\hat{h_i}^{(k+1)}$ using the Gerchberg-Saxton algorithm as described below.

(c) Compute the new PSF: $\hat{h_i}^{(k+1)}(x, y, z) = |TF_{2D}^{-1}\left(P_d(k_x, k_y, z)\, {P_{ai}}^{(k+1)}(k_x, k_y)\right)|^4$

Gerchberg-Saxton algorithm [4] allows the estimation of $P_{ai}^{(k+1)}$ by alternating constraints in the spatial domain (magnitude of h_{Ai} given by h_i) and in the frequency domain (the shape of h_{Ai} and the support of P_{ai} given by C) as follows:

(i) For each z-slice compute ${h_{Ai}}^{(k+1)}(x, y, z) = TF_{2D}^{-1}\left(P_d(k_x, k_y, z)\,.{P_{ai}}^{(k)}(k_x, k_y)\right)$

(ii) Replace the module of ${h_{Ai}}^{(k+1)}$ by $\sqrt[4]{\hat{h_i}^{(k+1)}}$

(iii) For each z-slice compute ${P_{aiz}}^{(k+1)}(k_x, k_y) = TF_{2D}\left({h_{Ai}}^{(k+1)}(x, y, z)\right) {P_d}^{-1}(k_x, k_y, z)$, then compute ${P_{ai}}^{(k+1)}$ as the average of these estimated aberration terms for different z-slices.

(iv) Project ${P_{ai}}^{(k+1)}$ in the set of function having their support included in $C = \{(k_x, k_y) \in \mathbb{N}^2; \sqrt{k_x^2 + k_y^2} < \frac{2\pi}{\lambda} NA\}$.

The convergence of the Gerchberg-Saxton was proved in [7]. The convergence of the proposed minimization algorithm under a set of constraints to a local minimum of J will be the goal of our future work. Numerical results show however a good behavior of such an estimation scheme.

5. Numerical experiments

We numerically generated a 3D image of a bead shell of thickness 400 nm and of internal diameter of 5 μm. The image size is of $140 \times 140 \times 70$ voxels, the voxel size being 50 nm along (Y, Y) axis and 140 nm along Z axis. An (Y, Z) slice of that image is shown in figure 1 (b). We degraded this image by a DV blur (for each z-slice, we used a new 3D PSF calculated by the model described in [3, 5]) and a Poisson noise, the background noise is set to $b_g = 10^{-4}$ (see figure 1 (a)). Note that a high value of the background constant decreases the signal to noise ratio. We then restored this image using a combination of 3 PSFs. The regularization parameters are set by try/error as follows: $\alpha = 10^{-3}, \beta_1 = 4.\,10^4, \beta_2 = 5.\,10^5, \beta_3 = 5.\,10^5$. The automatic estimation of these parameters will be the goal of our future work [8]. The comparison of the estimated image and one of 3 PSFs (figures 1 (c) and (f)) with those obtained by an algorithm without the pupil phase constraint [1] (figures 1 (d) and (g)) and comparison with the true image and PSF (see figures 1 (a) and (e)) show the advantage of the proposed approach. Our method allows to constrain the shape of the PSF. Note that the lobe of the original PSF is shifted from its central plane due to spherical aberration effects (see figure 1 (e)). Nevertheless, the estimated PSFs (figures 1 (f) and (g)) are centered because this shift is actually not estimated by the proposed method, even when considering the pupil phase constraints. An interesting future work is to estimate the PSF shift by introducing a strong constraint about this parameter. For example, one can obtain an information about the true object position (i.e. an information about the shift parameter) from a diffractive tomography microscopy system which can be coupled with the confocal microscopy system [9].

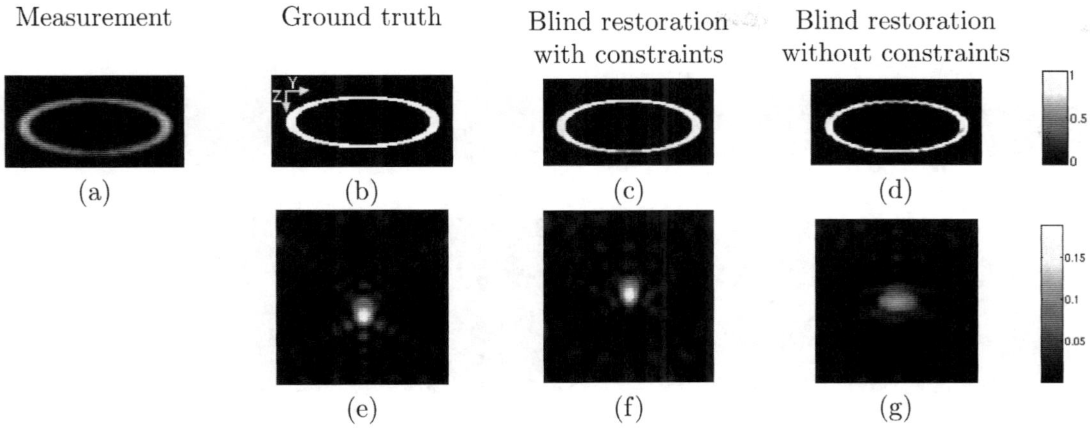

Figure 1. (Y, Z) of the (a) observation, the original image (b), the estimated image (c) and one of the 3 PSFs (f) using the proposed method, the estimated image (d) and one of the 3 PSFs (g) using the algorithm proposed in [1], the true PSF (e).

References

[1] S. Ben Hadj, L. Blanc-Fraud, G. Aubert, *Space Variant Blind Image Restoration.* Rapport de recherche INRIA 8073, September, 2012.

[2] S. Bonettini, R. Zanella, and L. Zanni. *A scaled gradient projection method for constrained image deblurring.* Inverse Problems, 2009, 25 : 015002.

[3] A. Egner, and S.W. Hell. *Aberrations in confocal and multi-photon fluorescence microscopy induced by refractive index mismatch.* Handbook of Biological Confocal Microscopy, 2006, 404–413.

[4] B.M. Hanser, M.G.L. Gustafsson, D.A. Agard, and J.W. Sedat. *Phase retrieval for high-numerical-aperture optical systems.* Optics letters, 2003, 28(10) : 801803.

[5] P.A. Stokseth. *Properties of a defocused optical system.* JOSA, 1969, 59(10) : 13141321.

[6] S. Yuan, and C. Preza. *3D fluorescence microscopy imaging accounting for depth-varying point-spread functions predicted by a strata interpolation method and a principal component analysis method.* SPIE, 2011, 19(23) : 23298-23314.

[7] J. R. Fienup. *Phase retrieval algorithms: a comparison.* Applied optics, 1982, 21(15) :2758–2769.

[8] X. Zhu, and P. Milanfar. *Automatic parameter selection for denoising algorithms using a no-reference measure of image content.* IEEE Transactions on Image Processing, 2010, 19(12) : 3116-3132.

[9] M. Sarmis, B. Simon, M. Debailleul, B. Colicchio, V. Georges, J.J. Delaunay, and O. Haeberle. *High resolution reection tomographic diffractive microscopy.* Journal of Modern Optics, 2010, 57(9) : 740-745.

Journal of Physics: Conference Series **464** (2013) 012010 doi:10.1088/1742-6596/464/1/012010

Reconstruction of 3D Radar Targets from Profile Functions in Arbitrary Directions with Level-set

Y Wen[1], N de Beaucoudrey[1], J Chauveau[2] and P. Pouliguen[3]
[1] IETR UMR 6164 CNRS, Université de Nantes, Lunam Université, Nantes, France
[2] DGA (Direction Générale de l'Armement), Maîtrise de l'Information, Bruz, France
[3] DGA (Direction Générale de l'Armement), DS/MRIS, Bagneux, France

E-mail: yan-hua.wen@etu.univ-nantes.fr, nicole.de-beaucoudrey@univ-nantes.fr

Abstract. The profile function of an object is defined as its transverse cross-sectional area versus distance along the observing direction and it is used by the ramp response technique to identify radar targets. Existing reconstruction algorithms have good performance with profile functions from 3 mutually orthogonal directions, while they give distorted results otherwise. To solve this inverse problem, we use the level set method, which iteratively deforms the shape of the target under a velocity field. An appropriate velocity is used and satisfactory reconstructed 3D images are presented for arbitrary directions.

1. Introduction

Radar imaging from profile functions using the ramp response technique [1] can be used for generating 3-dimensional images of radar targets (even stealthy or buried) so as to identify them. With existing reconstruction algorithms [2] [3], reconstructed 3D images are quite accurate for mutually orthogonal observing angles, while distorted results are obtained for non-orthogonal directions. Nevertheless, due to the limited view range of remote sensing radar equipments, in practice, it is almost impossible to guarantee 3 mutually orthogonal observing directions. Therefore, in this paper, we are concerned with image reconstruction from profile functions in arbitrary directions.
The level set method, devised by Osher and Sethian [4] [5], represents the evolving object as the zero level of an implicit higher dimensional function. This implicit representation keeps the object binary and evolves the indirect parameter, i.e. the level set function, instead of the parameterized shapes, which is completely adapted to our problem. Moreover, it has been effectively illustrated in the inverse scattering problem [6] [7]. Consequently, we apply this method to obtain an optimal shape of the target with profile functions in arbitrary directions.

2. Radar imaging from profile functions

As presented in [1], the ramp response of a radar target $h_r(t)$ is approximately proportional to its exact cross-sectional area perpendicular to the observing direction, noted the "geometrical" profile functions, $A_g(x)$:

$$h_r(t) \approx -\frac{1}{\pi c^2} A_g(x) \quad with \quad x = \frac{ct}{2} \tag{1}$$

where c is the speed of light in freespace, t the time variable, and x the space variable.

For applications and experiments using the ramp response technique, the profile function, defined as the "physical" profile function $A_p(x)$, which is an estimate of the geometrical profile function, is calculated from its transient ramp response by:

$$A_p(x) = -\pi c^2 h_r(t) \approx A_g(x) \quad with \quad x = \frac{ct}{2} \tag{2}$$

The initial reconstruction algorithm from profile functions proposed by Young [2], named "approximate limiting surface", uses a set of hyperbolic surfaces limiting the contour of the object for each of 3 observing directions and it is limited to single convex objects. On the contrary, the reconstruction algorithm presented in [3] uses the product of profile functions in 3 directions as a weighting function to account for the probability that one point belongs to the object. It exploits the information of profile functions more efficiently, and therefore extends to non-convex and separated objects. Unfortunately, both algorithms generate distorted results in non-orthogonal directions. Therefore, we aim to find a method for reconstruction in arbitrary directions.

3. Reconstruction using the level-set method

We attempt to use an iterative process to obtain an optimal estimate of the target by minimizing the mismatch between the data, i.e. profile functions of the unknown object, and the profile functions of the updated object. Our problem can be formalized in the form of

$$M^d u = A^d \tag{3}$$

where A^d is the observed data, the profile function observed in direction d, u the vector representing the unknown binary object we are looking for, M^d the observing matrix (or mapping matrix) and d the index number of directions. $d = 1, 2, 3$ in our case with 3 observing directions.

Most of iterative methods evolve the explicit function of object shape during the iteration. On the contrary, the level set method [4] represents the shape as the zero level of a higher order level set function φ, which is negative inside the object and positive outside, so as to keep the evolving object binary and to perform shape deformation in an implicit way.

The deformation of the object is formalized as a Hamilton-Jacobi equation for the level set function φ under a velocity field [7]:

$$\frac{\partial \varphi}{\partial t} + V|\nabla \varphi| = 0 \tag{4}$$

where φ is the level set function, $|\nabla\varphi|$ the gradient of φ, V the velocity.

Following the formulation in [7] and the desirable properties of the velocity [8], we propose a velocity adapted to our case:

$$V = -\sum_d (M^d)^T (M^d u - A^d) \tag{5}$$

where $(X)^T$ is the transpose of a matrix X.

4. Numerical results

In this section, illustrative results, which are reconstructed using the level set method, are presented. It is important to note that geometrical profile functions are used to validate this method. Hence, the mapping matrix M describes the relationship between the object and the geometrical profile function. The algorithm to construct the geometrical mapping matrix M and to calculate the geometrical profile function of a 3D object is presented in [9].

To show the performance in arbitrary directions, we choose 2 sets of observing directions: 3 mutually orthogonal directions ($d1 = x$ (90°, 0°), $d2 = y$, $d3 = z$) and 3 non-orthogonal directions ($d1$ = (90°,

75°), d2 = y, d3 = z). Actually, in [8], the implementation of the velocity was incorrect and it required a large number of memory space. Therefore, a very simple cubic object with as few as $N^3=16^3$ pixels was tested. Here, 4 cases of various objects are considered: a sphere, an asymmetric object, two separated objects (a sphere and a cone) and a square step-cylinder, which are shown in figure 1 (a) – (d), respectively. For each object, a cubic computational domain equal to 1.5 times the object maximum dimension is used. Each computational domain is distributed into $N^3=64^3$ cells.

To start the iterative process, an initial guess is required. It is somewhat arbitrary, only should be inside the computational domain. Therefore, for each object, we choose a small cube of 1/8 of the object dimension as the initial estimate. Once the initial contour is given, the initial level set function φ_0 is defined as the signed distance between each point and the contour, negative inside and positive outside [7]. The gradient of the level set function, $|\nabla\varphi|$, is approximated by a Hamiltonian scheme [7]. The velocity of the evolution is given by (5). At each iteration k, the normalized residual is chosen as:

$$r_k = \frac{1}{3}\sum_{d=1}^{3}\frac{\left\|A^d - M^d u_k\right\|_{L^2}}{\left\|A^d\right\|_{L^2}} \tag{6}$$

where $\|X\|_{L^2}$ is the Euclidean norm of X.

The residual decreases fast when a large time step δt is applied. But for the convergence and the stability of the Hamilton equation (4), the choice of the time step δt should satisfy the Courant–Friedrichs–Lewy condition (CFL condition) [10]:

$$\delta t \sum_i \frac{Vx_i}{\Delta x_i} \le C_{\max} \qquad i = 1 \cdots n \tag{7}$$

where n is 3 for three-dimensional case, Δx_i the spatial interval and $C_{max} = 1$ is typically used for an explicit problem.

To stop the iteration when an acceptable residual is achieved, we set the tolerance of the residual r_k to $\varepsilon = 10^{-2}$ and a maximum number of iterations $k_{\max} = 80$. To show the performance of level-set method, for each object in figure 1 (a) – (d), we compare the image reconstructed from orthogonal directions (ii) and that reconstructed from non-orthogonal directions (iii). As can be seen in figure 1 (ii), with geometrical profile functions from 3 mutually orthogonal directions, 3D images reconstructed by the level-set method are in good agreement with the original shapes. In the non-orthogonal case, figure 1 (iii), the two directions, *d1 = (90°, 75°)* and *d2 = y*, are very close to each other, with a small angle of 15°, therefore the reconstructed results are not as correct as those obtained in orthogonal case, but are still quite accurate to meet the need of identification. Comparing to the distorted images presented in [11], which are obtained by the algorithm of [3] from 3 non-orthogonal directions, the performance of reconstruction from profile functions is greatly improved. Although, the images reconstructed by the level-set method are not perfectly identical to the original shapes, they might be optimized by considering the regularization of the evolution, for example, smoothing contours.

The normalized residual r_k vs. the number of iterations k for both the orthogonal and non-orthogonal cases are compared in figure 2 for the corresponding objects in figure 1. For the objects in (a) – (c), the difference of the normalized residuals between the orthogonal and non-orthogonal cases are very small. The residuals decrease very rapidly during the first 20 iterations, and they remain stable when a certain small value is achieved. In figure 2, the residual value is given at the final iteration for the non-orthogonal case. Reconstructed images shown in figure 1 are obtained at the iteration $k_{\max} = 80$, except for the step cylinder (figure 2 (d)) in the non- orthogonal case, where the residual converges faster than that of the orthogonal case: Its evolution stops after 44 iterations with $r_k = 0.009$, which is less than the tolerance ε. Once again, figure 2 proves that reconstruction with the level-set method is sensitive neither to the relationship between the observing directions nor to the object shapes.

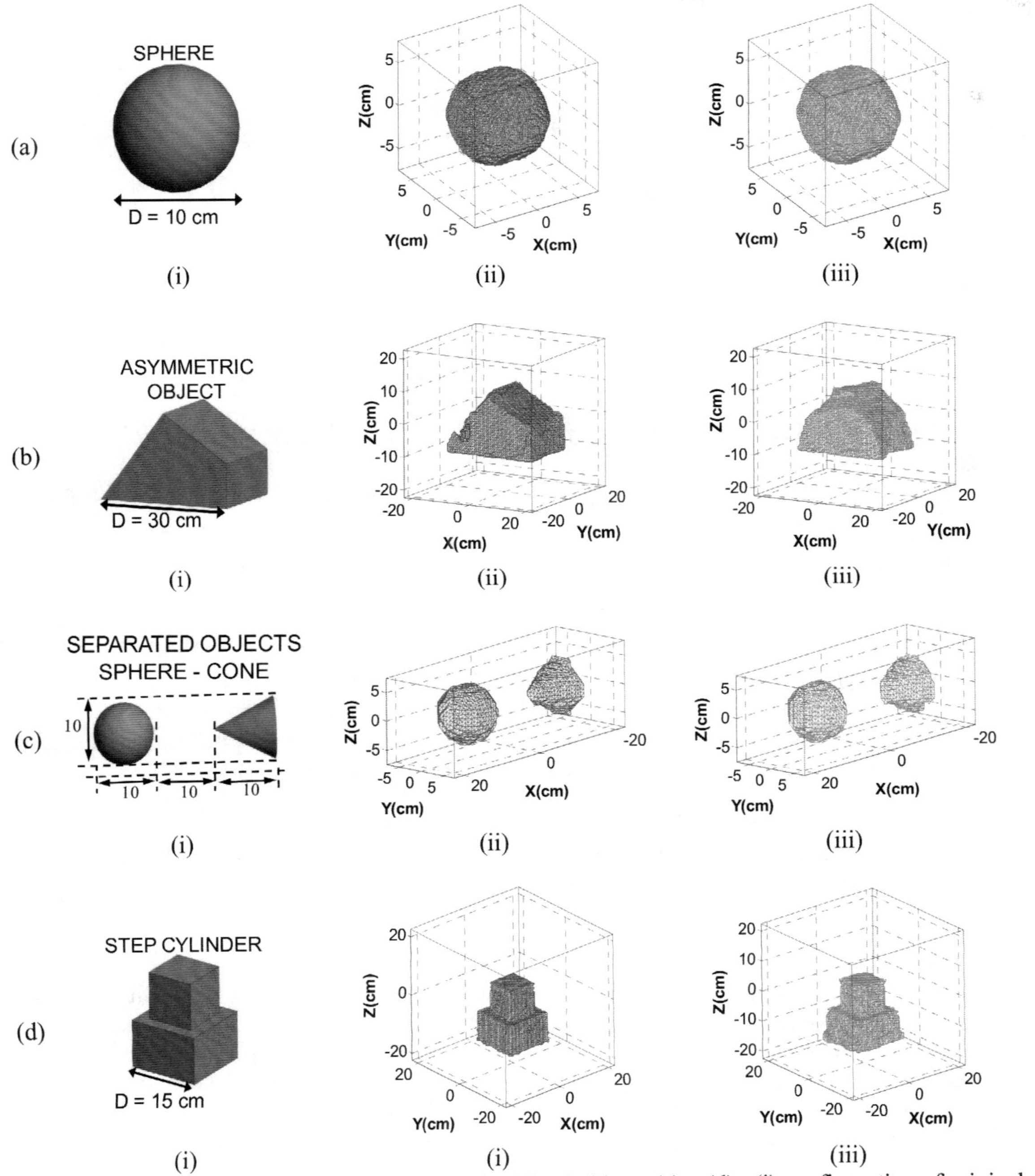

Figure 1. 3D reconstruction with level-set method for 4 objects (a) – (d) : (i) configuration of original object, (ii) reconstructed image obtained in orthogonal case, (iii) reconstructed image obtained in non-orthogonal case.

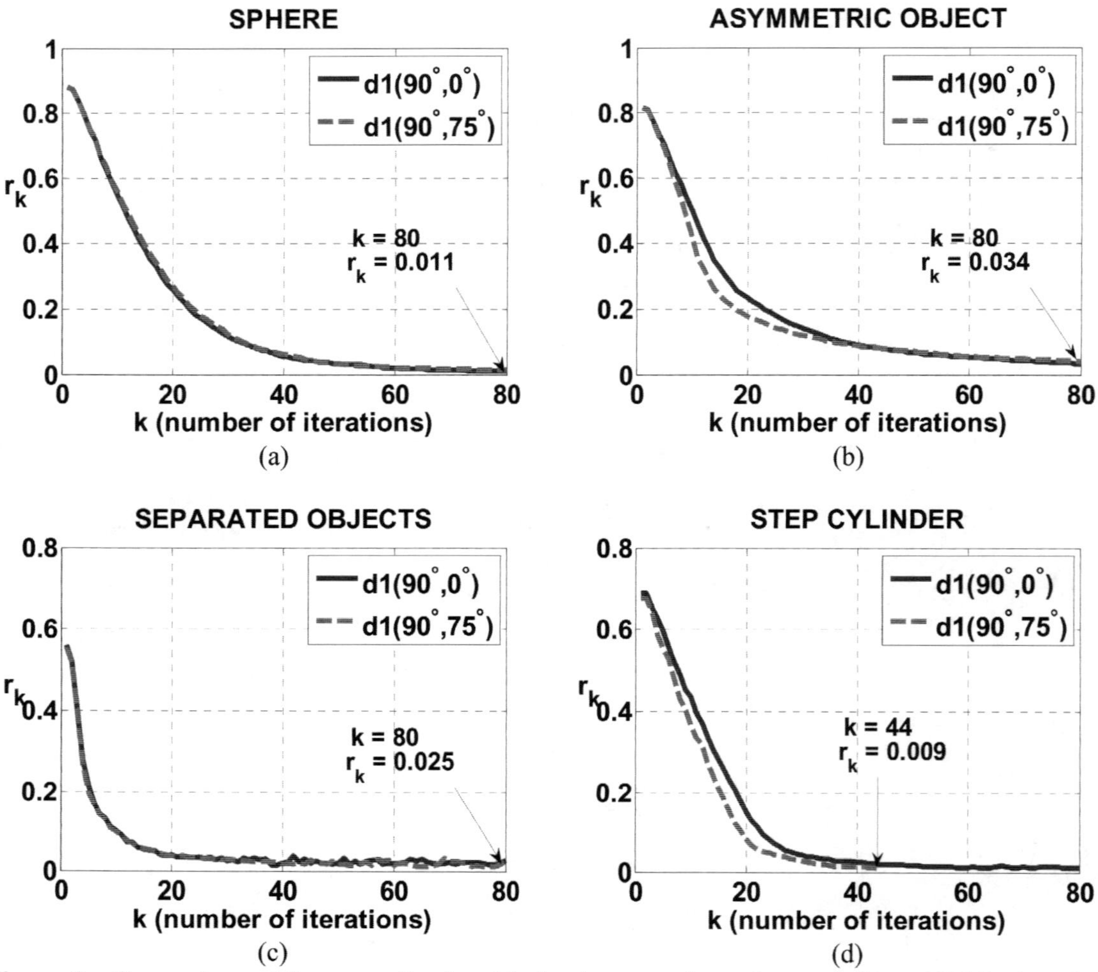

Figure 2. Comparison of the normalized residual r_k between the orthogonal (blue solid line) and non-orthogonal cases (red dash line) for the objects of figure 1.

5. Conclusion and perspectives

With the ramp response technique, radar imaging from profile functions only in 3 directions can produce an approximate contour for the target (even stealthy or buried). Previous algorithms for image reconstruction from profile functions work well in mutually orthogonal directions, but generate distorted estimates in non-orthogonal case. To optimize this performance, the level set method, which has been proved to be effective in shape optimization, is used. An adaptive velocity in our case is used and promising results of reconstruction for single and/or separated objects in arbitrary directions are obtained.

Until now, to validate the level-set method for reconstructions with profile functions, we only use the geometrical profile functions, namely the ideal case. The physical profile functions obtained from the real data will be considered. Moreover, to study the stability of the level-set method, some noise will be added in the data for the reconstruction process. The current mapping matrix we use is to depict the geometrical data of the original object and the estimate, therefore in the future, we are aim to find a mapping matrix adapted for the physical case.

Journal of Physics: Conference Series **464** (2013) 012010 doi:10.1088/1742-6596/464/1/012010

References

[1] Kennaugh E M and Moffatt D L 1965 Transient and impulse response approximations *Proc IEEE* 53 893-901
[2] Young J D 1976 Radar imaging from ramp response signatures *IEEE Trans. Antennas* AP-24 276-282
[3] Chauveau J and de Beaucoudrey N 2011 A new algorithm of 3D image reconstruction of radar targets from ramp responses in low frequency *Progress In Electromagnetics Research* M16 1-18
[4] Osher S and Sethian J 1988 A Front propagating with curvature-dependent speed: algorithms based on Hamiton-Jacobi formulations *J. Comput. Phys.* 7 912-49
[5] Sethian J A 1999 *Level set methods and fast marching methods* (Cambridge: Cambridge University Press)
[6] Litman A, Lesselier D and Santosa F 1998 Reconstruction of a two-dimensional binary obstacle by controlled evolution of a level-set *Inverse Problems* 14 685-706
[7] Santosa F 1996 A level set approach for inverse problems involving obstacles *Control Optimizat. Calculus Variat* 1 17-33
[8] Wen Y, de Beaucoudrey N, Saillard J, Chauveau J and Pouliguen P A level-set approach for radar imaging from ramp response in arbitrary directions *2nd Int. Workshop on New Computational Methods for Inverse Problems (NCMIP2012) Journal of Physics: Conference Series* 386 (2012)
[9] Wen Y, de Beaucoudrey N, Chauveau J and Pouliguen P 2013 Algorithm for profile function calculation of 3D objects: application for radar target identification in low frequency. *Progress In Electromagnetics Research* B50 273-293
[10] Burger M and Osher S L 2005 A Survey on Level Set Methods for Inverse Problems and Optimal Design. *European Journal of Applied Mathematics* 16 263-301
[11] Chauveau J, Wen Y-H and de Beaucoudrey N Low frequency radar target imaging using ramp response signatures in arbitrary directions *PIERS Proceedings, Marrakesh, MOROCCO, March* 20-23 1572 – 1575 (2011)

Journal of Physics: Conference Series **464** (2013) 012011 doi:10.1088/1742-6596/464/1/012011

A Lagrange method based L-curve for image restoration

G. Landi

Department of Mathematics, University of Bologna, Italy

E-mail: germana.landi@unibo.it

Abstract. The solution of image restoration problems usually requires the use of regularization strategies. The L-curve criterium is a popular heuristic tool for choosing good regularized solutions, when the data noise norm is not a priori known. In this work, we propose replacing the original image restoration problem with a noise-independent equality constrained one and solving it by an iterative Lagrange method. The sequence of the computed iterates defines a discrete L-shaped curve. By numerical results, we show that good regularized solutions correspond with the corner of this curve.

1. Introduction

Image restoration is an inverse problem usually modeled by the linear system

$$Ax + e = b \tag{1}$$

where $b \in \mathbb{R}^{n^2}$ represents the observed image, $e \in \mathbb{R}^{n^2}$ models additive noise ($\|b\| > \|e\|$), and $x \in \mathbb{R}^{n^2}$ is the original true image. The matrix A is a large-scale ill-conditioned matrix deriving from the discretization of a Fredholm integral equation. We consider the case of space-invariant blurring operators with periodic boundary conditions, so that A is block circulant with circulant blocks (BCCB) and matrix-vector products can be performed by using FFTs [1]. We suppose that A has full rank. Because of the ill-conditioning of A, regularization is necessary in order to reduce the sensitivity of the solution of (1) to the noise in b [2, 1].

The Conjugate Gradient method applied to the normal equations (CGNR) is a popular regularization technique [3] widely used to solve large-scale problems. The number of performed iterations is the regularization parameter. The L-curve criterium [4, 5] is an heuristic method for the regularization parameter choice which does not require any prior information about the error norm. This criterium is basically based on the plot, in a log-log scale, of the regularized solution norm versus the residual norm, for several regularization parameter values. Very often, this curve is L-shaped and its corner corresponds to a suitable value of the regularization parameter. In [6, 7], the so-called residual L-curve, the graph of the residual norm versus the iteration number, is proposed for computing the CGNR regularization parameter. The recent literature on inverse problems indicates a great interest in the development of heuristic regularization parameter choice rules, as the L-curve-based ones, since they can be used in real-world applications where no information about the error norm is usually available.

The main contribution of this work consists in proposing a new heuristic criterium, referred to as *Lagrangian L-curve* criterium, for computing regularized solutions of (1). The proposed criterium is well-suited for image restoration applications where the coefficient matrix is available only as an operator for matrix-vector products. A wide numerical experimentation has been performed in order to assess the effectiveness of the proposed approach by numerical evidence.

The sequel is organized as follows. The new Lagrangian L-curve criterium is described in section 2 and the numerical results are presented in section 3; conclusions are given in section 4.

2. The Lagrangian L-curve

Consider the following noise-independent minimization problem

$$\begin{aligned} \text{minimize} \quad & \frac{1}{2}\|x\|^2 \\ \text{subject to} \quad & h(x) = 0 \end{aligned} \tag{2}$$

where $h(x) = \frac{1}{2}\|Ax - b\|^2 - \varepsilon$ and ε is a small positive parameter introduced to overcome the non-regularity of the minimizer of $\|x\|^2$ subject to $\|Ax - b\|^2 = 0$. (Here and in the following, $\|\cdot\|$ denotes the Euclidean norm.) It should be chosen so that $\varepsilon \ll \rho^2/2$, where $\rho = \|e\|$ is the data noise norm. Even under the hypothesis of no knowledge about the noise norm, this is not a restrictive condition since, in practice, ε can be chosen close to the machine precision.

Problem (2) has been introduced in [8] for the regularization of discrete ill-posed problems when the noise norm is supposed to be known. In this work, we suppose the noise norm to be not explicitly known and propose an heuristic regularization strategy.

The first-order optimality conditions for (2) are

$$\begin{cases} \nabla_x \mathcal{L}(x, \lambda) = 0 \\ h(x) = 0 \end{cases} \tag{3}$$

where $\mathcal{L}(x, \lambda) = \frac{1}{2}\|x\|^2 + \lambda h(x)$ is the Lagrangian function and $\lambda \in \mathbb{R}$ is the Lagrange multiplier. Problem (2) is equivalent to the strictly convex problem of minimizing $1/2\|x\|^2$ over the convex set $h(x) \leq 0$, then its solution is unique with positive Lagrange multiplier [9, 10].

Given an initial iterate (x_0, λ_0), an iteration of the Lagrange method to solve (2) is stated as

$$\begin{aligned} x_{k+1} &= x_k + \alpha_k \Delta x_k \\ \lambda_{k+1} &= \lambda_k + \alpha_k \Delta \lambda_k \end{aligned} \tag{4}$$

where $(\Delta x_k, \Delta \lambda_k)$ is the Newton step solving the linear system

$$\begin{pmatrix} \nabla^2_{xx}\mathcal{L}(x_k, \lambda_k) & \nabla_x h(x_k) \\ \nabla_x h(x_k)^T & 0 \end{pmatrix} \begin{pmatrix} \Delta x \\ \Delta \lambda \end{pmatrix} = - \begin{pmatrix} \nabla_x \mathcal{L}(x_k, \lambda_k) \\ h(x_k) \end{pmatrix} \tag{5}$$

and the step-length α_k is chosen in order to guarantee a sufficient decrease of the merit function

$$m(x, \lambda) = \frac{1}{2}\Big(\|\nabla_x \mathcal{L}(x, \lambda)\|^2 + |h(x)|^2\Big). \tag{6}$$

In this work, we make use of the Armijo rule for the selection of the step-length α_k [11]. Under standard assumptions for the Lagrange method convergence, the sequence $\{(x_k, \lambda_k)\}$ converges to a point satisfying the first-order necessary conditions for a solution to (2).

Let $r_k = b - Ax_k$ be the residual at the k-th iterate. The following lemma show that, even if the sequences $\{\|x_k\|\}_\mathbb{N}$ and $\{\|r_k\|\}_\mathbb{N}$ do not have a monotonic behavior, an eventually finite non empty index subset $\mathcal{K}$ exists such that $\{\|x_k\|\}_\mathcal{K}$ and $\{\|r_k\|\}_\mathcal{K}$ are respectively increasing and decreasing with $k \in \mathcal{K}$.

Lemma 2.1. *Let x^* be a minimum point of (2) and let λ^* be the corresponding Lagrange multiplier. If the sequence $\{(x_k, \lambda_k)\}_{\mathbb{N}}$ generated by the iterative procedure (4) converges to (x^*, λ^*), then, given $x_0 = 0$ and $\varepsilon \ll \rho^2/2$, the solution norm sequence $\{\|x_k\|\}_{\mathbb{N}}$ and the residual norm sequence $\{\|r_k\|\}_{\mathbb{N}}$ respectively converge to $\|x^*\|$ and $\|r^*\|$ such that*

$$\|x^*\| > \|x_0\| = 0, \tag{7}$$
$$\|r^*\| < \|r_0\|. \tag{8}$$

Moreover, a non empty, eventually finite, index subset $\mathcal{K}$ exists such that $\{\|x_k\|\}_{\mathcal{K}}$ and $\{\|r_k\|\}_{\mathcal{K}}$ are respectively increasing and decreasing with $k \in \mathcal{K}$.

Proof. The initial guess x_0 is the solution of the unconstrained problem of minimizing $\|x\|^2$; however, x_0 is not the solution of (2) since $\|r_0\|^2 = \|b\|^2 \geq \rho^2 > 2\varepsilon$. Thus, $x^* \neq x_0$ and $\|r_0\| > \sqrt{2\varepsilon} = \|r^*\|$, $\|x^*\| > \|x_0\|$. Let us now consider the sequences $a_k := \|x_0\| - \|x_k\|$ and $b_k := \|r_0\| - \|r_k\|$. They respectively converge to $a^* = \|x_0\| - \|x^*\| < 0$ and $b^* = \|r_0\| - \|r^*\| > 0$. Then, two positive integers $\overline{k}_1 \in \mathbb{N}$ and $\overline{k}_2 \in \mathbb{N}$ exist such that $a_k < 0$ for all $k \geq \overline{k}_1$ and $b_k > 0$ for all $k \geq \overline{k}_2$. By setting $\overline{k} = \max\{\overline{k}_1, \overline{k}_2\}$ we get $a_k < 0$ and $b_k > 0$ for all $k \geq \overline{k}$. Moreover, we have $\|x_0\| < \|x_{\overline{k}}\|$ and $\|r_0\| > \|r_{\overline{k}}\|$ for all $k \geq \overline{k}$. This shows that a non empty index subset $\mathcal{K}$ always exists and, in the worst case, it reduces to $\{0, \overline{k}\}$. □

Let now $\mathcal{K}$ be an index subset such that $\{\|x_k\|\}_{\mathcal{K}}$ and $\{\|r_k\|\}_{\mathcal{K}}$ are respectively increasing and decreasing with $k \in \mathcal{K}$. Finally, consider the two-dimensional graph of the points

$$\Big(\log_{10}\|Ax_k - b\|, \log_{10}\|x_k\|\Big), \quad k \in \mathcal{K}. \tag{9}$$

We will refer to it as the *Lagrangian L-curve* since it is based on iterations of the Lagrange method for (2) and since, very often, this graph looks like the letter L. We propose to choose the iterate x_k corresponding to the vertex point of the Lagrangian L-curve as an approximated solution to (1).

Basically, the proposed Lagrangian L-curve criterium is used to single out one of the Lagrange iterates that may be a good regularized solution, even if the limit of the entire Lagrange sequence is not a useful one. A justification for why one of the Lagrange iterate can be used as a regularized solution can be given as follows. Given $x_0 = 0$, the sequence x_k converges to the solution of (2) and a finite index ℓ exists such that $\|Ax_\ell - b\| < \rho$. Such an iterate x_ℓ, satisfies the *discrepancy principle* [12] and, as shown in [8], is a useful regularized solution. An extensive numerical experimentation has shown that x_ℓ is close to the Lagrangian L-curve corner.

In this paper, we present numerical evidence that the Lagrangian L-curve criterium can be used to determine good regularized solution to (1).

3. Implementation details

Initial choice λ_0. The Lagrange method requires an initial guess for λ_0. As suggested in §14.4 of [11], it can be chosen such that

$$\lambda_0 = \Big(\nabla_x h(x_0)^T \nabla_x h(x_0)\Big)^{-1}\Big[h(x_0) - \nabla_x h(x_0)^T x_0\Big]. \tag{10}$$

Computation of the search direction $(\Delta x_k, \Delta\lambda_k)$. When A is BCCB, $\nabla^2_{xx}\mathcal{L}(x_k, \lambda_k)$ is BCCB again and can be easily inverted by means of FFTs. Hence, an explicit solution of (5) can be computed as (§14.1 of [11]):

$$\begin{aligned}\Delta\lambda_k &= -\Big(D(x_k, \lambda_k)\Big)^{-1}\Big(\nabla_x h(x_k)^T(\nabla^2_{xx}\mathcal{L}(x_k, \lambda_k))^{-1}\nabla_x\mathcal{L}(x_k, \lambda_k) - h(x_k)\Big)\\ \Delta x_k &= -\Big(\nabla^2_{xx}\mathcal{L}(x_k, \lambda_k)\Big)^{-1}\Big(\nabla_x h(x_k)\Delta\lambda_k + \nabla_x\mathcal{L}(x_k, \lambda_k)\Big)\end{aligned} \tag{11}$$

where $D(x_k, \lambda_k) = \nabla_x h(x_k)^T \left(\nabla^2_{xx}\mathcal{L}(x_k, \lambda_k)\right)^{-1}\nabla_x h(x_k)$ is a scalar.

Computation of the index subset $\mathcal{K}$. In order to create the index subset $\mathcal{K}$, we propose to check the monotonicity of the solution and residual norm sequences and discard those points where the monotonicity condition is not satisfied. In our implementation, the algorithm for the index subset $\mathcal{K}$ computation starts setting $0 \in \mathcal{K}$. Given $k \in \mathcal{K}$, the term in $\mathcal{K}$ which follows k is the least index $\phi(k) \in \mathbb{N}$ such that $\|x_{\phi(k)}\| > \|x_k\|$ and $\|r_{\phi(k)}\| < \|r_k\|$.

Stopping criteria. In our implementation, the iterative procedure (4) has been terminated when the relative distance between two successive iterates (x_k, λ_k) has became smaller than a given positive tolerance τ or when a maximum number of allowed iterations has been performed.

4. Numerical results

The numerical experiments have been executed on a Sun Fire V40z server consisting of four 2.4GHz AMD Opteron 850 processors with 16GB RAM using Matlab 7.10 (Release R2010a). For all the tests, a maximum number of 30 iterations and the value $\tau = 10^{-8}$ have been set for terminating the Lagrange iteration. The value of the parameter ε has been set equal to the machine precision. The proposed Lagrangian L-curve criterium has been compared with the traditional L-curve criterium [4, 5] and the residual L-curve criterium [7] for the CGLS method. In all cases, the L-corner method [13] has been used for the detection of the curves corner. A number of 150 CGLS iterations have been performed in order to determine the corresponding L-curve and residual L-curve.

The test problems are based on 9 famous 256×256 test images: the *Lena*, *Cameraman*, *Mandril*, *Peppers*, *Jetplane*, *Walkbridge*, *Woman*, *Boat* and *Lake* images which can be obtained from `http://decsai.ugr.es/cvg/dbimagenes/g256.php`. These images have been corrupted by Gaussian blur with variance $\sigma = 2$ and 4, and defocus blur with radius $R = 7.5$ and 15, respectively obtained with the code `psfGauss` and `psfDefocus` from [1]. Then Gaussian noise has been added with several noise level values. (The noise level NL is defined as NL $:= \|e\|/\|Ax_{\text{exact}}\|$ where $\mathbf{x}_{\text{exact}}$ is the exact image.) More specifically, the following values for the noise level have been used: NL$=5{\cdot}10^{-4}, 7.5{\cdot}10^{-4}, 1{\cdot}10^{-3}, 2.5{\cdot}10^{-3}, 5{\cdot}10^{-3}, 7.5{\cdot}10^{-3}, 1{\cdot}10^{-2}, 2.5{\cdot}10^{-2}, 5{\cdot}10^{-2}$. For each noise level, 10 Gaussian noise vectors have been obtained with different noise realizations. This amounts to 360 experiments for each image and to 3240 experiments in total.

For all the considered criteria (Lagrangian L-curve, L-curve and residual L-curve), let k_{corner} and k_{best} denote the iteration corresponding respectively to the L-curve corner and to the smallest relative error. The performance of the three criteria has been evaluated in terms of percentage of experiments producing a reconstructed image such that

$$\|x_{k_{\text{corner}}} - x_{\text{exact}}\| > C\|x_{k_{\text{best}}} - x_{\text{exact}}\|, \quad \text{with} \quad C = 1.05, 1.25, 2.5. \tag{12}$$

For each image, the corresponding percentage values are reported in table 1; for the total 3240 experiments, the percentage values are also shown. The numerical results in the table indicate that, in all the experiments, the proposed criterium outperforms the other ones. Figure 1 depicts the data images, corresponding to Gaussian blur with $\sigma = 2$ and Gaussian noise with NL$=2.5{\cdot}10^{-3}$; figure 2 shows the corresponding reconstructions obtained by the Lagrangian L-curve criterium. For the *Lena* image degraded by Gaussian blur and Gaussian noise ($\sigma = 2$, NL$=10^{-2}$), figure 3 represents the reconstructions obtained by the Lagrangian L-curve, the L-curve and the residual L-curve criteria. The L-curve image tends to be under-regularized while the Residual L-curve image tends to be over-regularized. The same behavior of the L-curve and the Residual L-curve criteria have been observed in almost all the experiments. Figure 4 shows the relative error behavior for the Lagrange and GCLS iterations together with the graph of the Lagrangian L-curve, L-curve and residual L-curve. Observe that, for the Lagrange iterations, the relative error grows rather slowly after the iteration with the smallest relative error even if

Journal of Physics: Conference Series **464** (2013) 012011 doi:10.1088/1742-6596/464/1/012011

Image	Lagr. L-curve			L-curve			Res. L-curve		
	> 1.05	> 1.25	> 2.5	> 1.05	> 1.25	> 2.5	> 1.05	> 1.25	Z2.5
Lena	25.56	8.33	0.00	60.83	42.50	6.94	88.89	61.11	0.00
Cameraman	9.17	5.56	0.00	58.06	50.00	10.56	84.72	36.11	0.00
Mandril	8.61	5.56	0.00	67.22	45.56	5.83	63.89	17.22	5.56
Peppers	15.83	5.56	0.00	60.00	48.89	5.56	86.94	31.39	0.83
Jetplane	17.50	9.44	2.50	81.11	58.33	22.22	83.89	38.33	6.11
Walkbridge	11.94	10.28	0.00	52.78	33.61	5.28	80.56	22.22	0.00
Woman	31.11	17.78	6.11	97.22	85.83	47.22	76.94	34.44	10.56
Boat	16.94	8.33	0.00	64.44	50.56	12.22	84.44	37.22	1.11
Lake	16.67	10.00	0.00	57.22	45.00	7.78	86.39	44.72	0.28
Total	17.04	8.98	0.96	66.54	51.14	13.73	81.85	35.86	2.72

Table 1: Numerical results.

ε is set to a value equal to the machine precision. This behavior of the relative error has been observed in all the performed numerical tests and is indeed a desirable property for a numerical technique aimed to solve ill-posed inverse problems. In fact, when the relative error grows slowly, then performing too many iterations does not dramatically deteriorate the reconstructed image quality.

5. Conclusions

In this work, a new heuristic criterium, referred to as Lagrangian L-curve criterium, has been presented for the solution of image restoration problems in the absence of prior information about the data noise norm. The effectiveness of the proposed criterium is shown numerically by performing numerous test problems on several images with various blur operators and noise level values. The numerical results are promising and indicate that the proposed criterium can be more effective than other state-of-the-art criteria for the CGLS method.

Figure 1: Data images (Gaussian blur $\sigma = 2$, Gaussian noise NL=$2.5 \cdot 10^{-3}$). Top to bottom, left to right: *Lena*, *Cameraman*, *Mandril*, *Peppers*, *Jetplane*, *Walkbridge*, *Woman*, *Boat*, *Lake*.

Journal of Physics: Conference Series **464** (2013) 012011 doi:10.1088/1742-6596/464/1/012011

Figure 2: Images restored by the Lagrangian L-curve criterium.

Figure 3: *Lena* data image ($\sigma = 2$, NL=10^{-2}). Top to bottom, left to right: data, Lagr. L-curve, L-curve, Res. L-curve images.

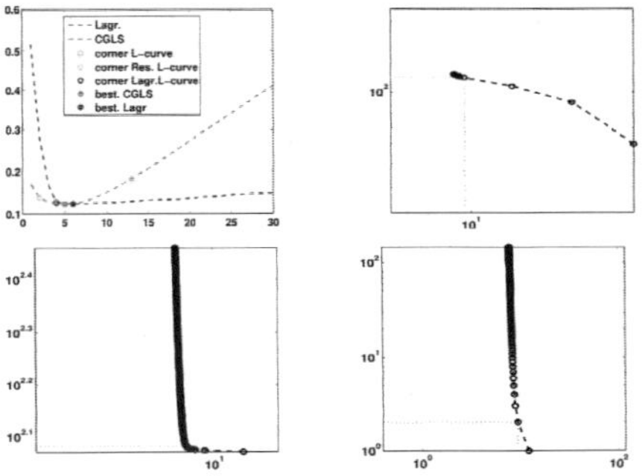

Figure 4: *Lena* image ($\sigma = 2$, NL=10^{-2}). Top to bottom, left to right: relative error (the circles and asterisks indicate the reconstruction and best relative error values), Lagr. L-curve, L-curve, Res. L-curve (residual norm versus solution norm).

[1] Hansen P C, Nagy J and O'Leary D P 2006 *Deblurring images. Matrices, Spectra and Filtering* (SIAM)
[2] Engl H W, Hanke M and Neubauer A 1996 *Regularization of inverse problems* (Kluwer Academic Publishers, Dordrecht)
[3] Hanke M 1995 *Conjugate Gradient Type Methods for Ill-Posed Problems* (Pitman Research Notes in Mathematics, Longman House, Harlow, Essex)
[4] Hansen P and O'Leary D 1993 *SIAM J. Sci. Comput.* **14** 1487–1503
[5] Hansen P C 1992 *SIAM Rev.* **34** 561–580
[6] Reichel L and Sadok H 2008 *J. Comput. Appl. Math.* **219** 493–508
[7] Reichel L and Rodriguez G 2012 *Num. Alg.* DOI 10.1007/s11075-012-9612-8
[8] Landi G and Piccolomini E L 2010 *Comput. Math. App* **60** 1723–1738
[9] Bertsekas D 2003 *Convex Analysis and Optimization* (Athena Scientific)
[10] Landi G 2008 *Comput. Optim. Appl.* **39** 347–368
[11] Luenberger D G 2003 *Linear and nonlinear programming* (Kluwer Academic Publishers, (2nd Ed.))
[12] Morozov V A 1984 *Methods for solving incorrectly posed problems* (Springer-Verlag, New York)
[13] Hansen P, Jensen T and Rodriguez G 2006 *J. Comput. Appl. Math.* **198** 483492

Journal of Physics: Conference Series **464** (2013) 012012 doi:10.1088/1742-6596/464/1/012012

Nonlinear mechanical properties estimated from images of a mechanical test

Florent Mathieu, Hugo Leclerc, François Hild, Stéphane Roux
LMT-Cachan, 61 avenue du Président Wilson, Cachan, France
E-mail: leclerc@lmt.ens-cachan.fr

Abstract. The problem of identifying the (non-linear) mechanical properties of a solid from images taken in a mechanical testing machine is considered. The two classical steps of i) measuring the displacement field with digital image (and volume) correlation and ii) identifying the mechanical properties from the displacement field are combined into a single "integrated" procedure to reduce the artificial bias introduced from the projection of the kinematic data onto an arbitrary displacement basis. It allows for the identification of nonlinear constitutive law parameters with arbitrarily complex kinematics. The example of the plastic behavior of thin Ti-foils is considered.

1. Introduction

1.1. Digital image and volume correlation

Due to the large amount of data it permits to extract, the digital image correlation (DIC) tools are now commonly used in the field of experimental mechanics. Given two images f and g representing e.g., the gray level or the density of the same part in two different mechanical loading conditions, DIC aims at finding how one of the two images is to be deformed to match the other one. In mechanical terms, the corresponding displacement field is an important piece of information for qualitative or quantitative observations.

Due to noise produced by usual acquisition devices, it is very common to obtain the displacement $\boldsymbol{u}(\boldsymbol{x})$ from the minimization of the following quadratic difference

$$\int_{\Omega} [g(\boldsymbol{x} + \boldsymbol{u}(\boldsymbol{x})) - f(\boldsymbol{x})]^2 \, \mathrm{d}\boldsymbol{x} \tag{1}$$

where Ω is the Region Of Interest (ROI).

Whether f contains a lot of contrast or not, Eq. (1) does not contain enough constraints to find independent values of $\boldsymbol{u}$ for each pixel or voxel of Ω. To circumvent this difficulty, in its earlier developments DIC consisted in registering small zones of interest or ZOIs (i.e., small interrogation windows [1]) in the considered ROI. Each registration were performed over the ZOI area with various kinematic hypotheses, possibly accounting for their warping. Global approaches, which appeared more recently [2, 3, 4], consist in performing a registration over the whole ROI. The kinematic bases are then defined over the whole ROI (i.e., Rayleigh-Ritz formulations) or over a discretized ROI (i.e., Galerkin formulations). In this context, a finite element formulation can be considered since it gives a natural link with numerical simulations of mechanical problems for which the very same discretization may be considered.

1.2. Finite Element Model Updating

DIC is a natural tool to provide the input data for Finite Element Model Updating (FEMU) procedures for identifying the mechanical properties of a solid which are consistent with the observed displacement. The principle of FEMU [5] is to tune the parameters of a constitutive law so as to minimize the distance between the "experimental" fields (obtained by DIC) and the numerical ones obtained from a direct simulation.

However, it is to be emphasized that the measured displacements are always corrupted by noise. Thus, to minimize the noise sensitivity, a "good" metric should account for this noise and weighting is definitely not uniform and uncorrelated over the whole ROI. When the same discretization can be used for DIC (providing a vector of measured degrees of freedom $\{\boldsymbol{u}_e\}$, and a correlation matrix $[\boldsymbol{M}]$) and for the numerical simulations (providing a vector $\{\boldsymbol{u}_n\}$), is has been proven [6] that

$$\langle \{\boldsymbol{u}_e - \boldsymbol{u}_n\}, [\boldsymbol{M}]\{\boldsymbol{u}_e - \boldsymbol{u}_n\}\rangle \tag{2}$$

is the optimal choice for the distance.

Besides, as finite element simulations require boundary conditions (at least on non free borders), in most published work, they come from the "experimental" DIC values, which in turn may be highly sensitive to noise. However, it is possible to consider that the boundary conditions are additional unknowns of the FEMU procedure. In such cases, the boundary values given by DIC are simply ignored (or used as an initial value of the procedure).

1.3. Integrated DIC (IDIC)

The previous ingredients have been used in different cases where it has given full satisfaction even for situations in which the Signal to Noise Ratio (SNR) is small (i.e., in elasticity [7]).

However, in this two step DIC / FEMU procedure:

- the elements in the mesh have to remain big enough because of the aforementioned DIC limitations. In the case of complex kinematics (e.g., with singularities), such big elements may not be able to capture the mechanical reality requiring fine discretizations;
- in extreme cases, due to noise, the DIC solutions may be stuck in local minima, even if they do not correspond to the solution of a mechanical problem (typically the one that are used for the FEMU procedure).

Starting from the observation that the DIC fields are only intermediate results (they are generally not the final goal), an integrated DIC procedure has been developed. In this procedure, Eq. (1) is still to be minimized, but $\boldsymbol{u}$ has to be the solution to the finite element problem. Thus, Eq. (1) is minimized with respect to the unknown mechanical properties (and the boundary conditions). This procedure is presented in Figure 1. In practice, the finite element solutions are (automatically and symbolically) differentiated and then injected into Eq. (1) to obtain the updates to be applied to the sought parameters.

The basis of this procedure has been presented in [6] but for a linear test case, and without considering the boundary conditions in the unknown set. In the following, some results are presented with this integrated DIC procedure in the case of a nonlinear constitutive law.

2. Experimental setup

In the present example, a specimen made of pure titanium (T35 grade) has been photographed for different load levels in a standard hydraulic testing machine. The test is displacement-controlled at a constant speed, and images are taken at regular intervals of time. As can be seen in Figure 2, the specimen has a "dogbone" shape. Although it is unusual to use such irregular shapes for testing, DIC allows to make use of this type of geometry, and it gives the advantage

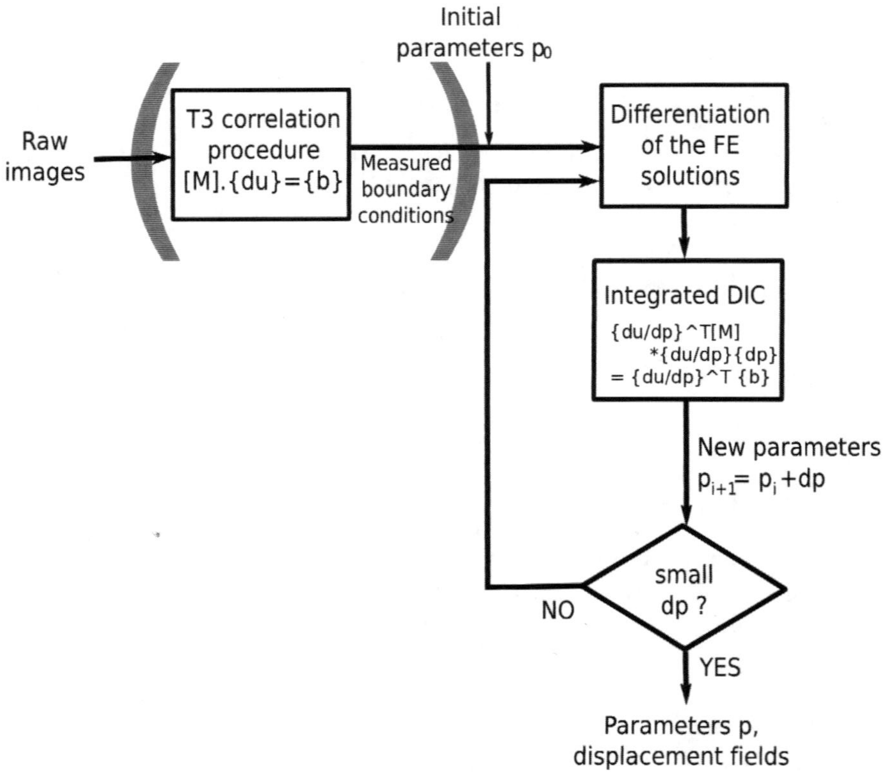

Figure 1. Integrated procedure for the identification of the parameters p_i. These parameters can be the constitutive law parameters, as in the present case, or boundary conditions. The initial correlation procedure, also with T3 (linear triangular) elements, can be used to initialize the identification procedure with known displacements at non-free boundaries (see Figure 2), but is not mandatory.

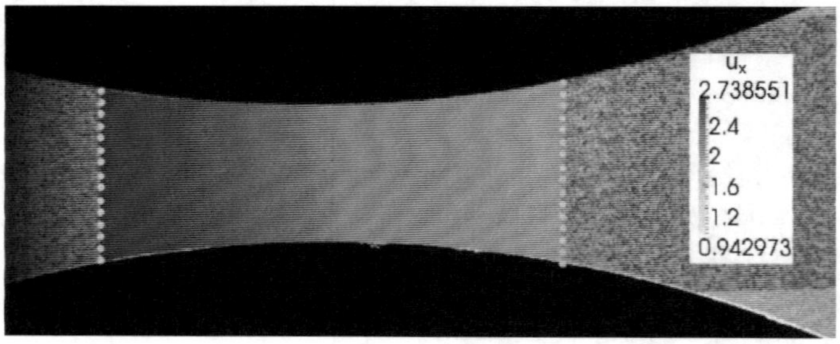

Figure 2. Specimen with the identified longitudinal $\boldsymbol{u}_x$ displacement field for image 10. The displacement is expressed in pixels (1 pixel $\leftrightarrow$ 27 μm). The green dots indicate the non free boundaries.

of providing several loading conditions on the same sample. Besides, it helps to ensure that the localization phenomena will occur in a given zone within the ROI.

Starting from macroscopic observations (i.e., from stress-strain curves), the nonlinear Ramberg-Osgood [8] expression has been chosen as a simple constitutive law for the simulation.

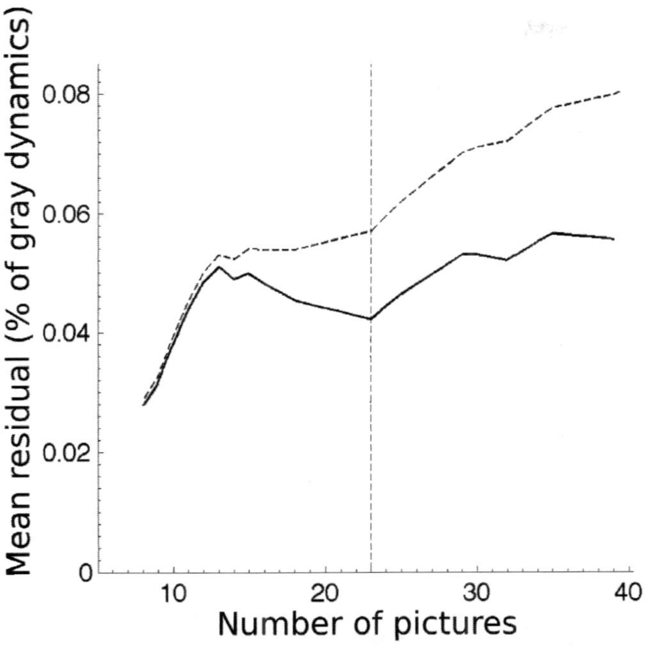

Figure 3. Residuals, obtained by FEMU (dashed lines) and by IDIC (solid ones). The gray level dynamics is the amplitude of observable gray level in the reference image.

Schematically, such a law describes the plastic strain as a power-law of the (deviatoric) stress. Only two parameters, namely the yield stress σ_0 for 0.2 % strain offset, and the hardening exponent n, are to be identified to define completely the constitutive law.

3. Exploitation

Figure 3 shows the gray level residuals (i.e., the rms difference between $f(\boldsymbol{x})$ and $g(\boldsymbol{x} + \boldsymbol{u}(\boldsymbol{x}))$ at the end of the minimization procedure) for a given number of images (the contribution of each image pair is summed to obtain the same parameters across the interval). In the present case, the constitutive law has been implemented such as automatic symbolic differentiation was still possible for all unknown parameters. The accuracy problems that arise from numerical differentiation procedures has thus been avoided.

For the two cases (FEMU and IDIC), the values of σ_0 and n are not completely stable. This means that the choice of the Ramberg-Osgood algebraic form is a too simple assumption. The important point to note however is that the residuals obtained by IDIC are significantly lower than the ones obtained by FEMU. The goal of developing IDIC procedures is to lower the gray level residuals, whereas that of FEMU is to reduce the distance to an intermediate result, namely, the measured displacement field. Hence, the final levels are different (the acquisition noise in terms of gray level was less than 1%). The automatic symbolic differentiation also allows to avoid the computation of as many finite element solutions as the number of parameters sought, and thus make this method less costly. Without this manipulation, the computation time is very similar to the one obtained with FEMU.

4. Conclusion

An integrated DIC procedure has been introduced to identify the parameters of a nonlinear constitutive law. In terms of fulfilment of the sought objective, the final results are clearly better than the ones obtained by the classical FEMU procedure. Furthermore, the proposed regularization allows the fields to be identified to better guide the numerical simulations in the bulk, notably for the cases where the boundary conditions do not give enough information.

References

[1] Sutton MA, Wolters WJ, Peters WH, Ranson W and McNeill SR. Determination of Displacements Using an Improved Digital Correlation Method *Im. Vis. Comp.* 1983;**1**:133-139.

[2] Broggiato GB. Adaptive image correlation technique for full-field strain measurement In Proc. 12th Int. Conf. Exp. Mech.; 2004.

[3] Sun Y, Pang J, Wong C and Su F. Finite-element formulation for a digital image correlation method. *Appl. Optics* 2005;**44**:7357-7363.

[4] Besnard G, Hild F and Roux S. "Finite-element" displacement fields analysis from digital images: Application to Portevin-Le Châtelier bands *Exp. Mech.* 2006;**46**:789-803.

[5] Kavanagh KT and Clough RW. Finite Element Applications in the Characterization of Elastic Solids *Int. J. Solids Struct.* 1971;**7**:11-23.

[6] Leclerc H, Périé J-N, Roux S and Hild F. Integrated digital image correlation for the identification of mechanical properties In: Proc. MIRAGE 2009. Springer; 2009,161-171.

[7] Gras R, Leclerc H, Roux S, Otin S, Schneider J and Périé J-N. Experimental Mechanics October 2012 Identification of the Out-of-Plane Shear Modulus of a 3D Woven Composite. *Exp. Mech.* 2013;in press (DOI:10.1007/s11340-012-9683-4).

[8] Ramberg W and Osgood W.R, Description of stress-strain curves by three parameters, in *Technical notes, National Advisory Committee for Aeronautics* 4th edition, 1943.

AUTHOR INDEX

Institute of Physics
Dirac House, Temple Back
Bristol BS1 6BE UK

CURRAN ASSOCIATES INC
proceedings.com